COAL IN
APPALACHIA

COAL IN APPALACHIA

An Economic Analysis

CURTIS E. HARVEY

THE UNIVERSITY PRESS OF KENTUCKY

Library of Congress Cataloging-in-Publication Data

Harvey, Curtis E.
  Coal in Appalachia.

  Bibliography: p.
  Includes index.
  1. Coal trade—Appalachian Region. I. Title.
HD9547.A127H37 1986   333.8′22′0974   86-13294
ISBN 0-8131-1577-9

309653

CONTENTS

PREFACE

Few topics captured national attention more in the past decade than energy. Although the urgency of this issue in the 1980s is no longer as great as in the 1970s, an awareness of its importance endures.

Of the energy-supplying sectors in the United States, coal is surely the most volatile. Plagued by an unenviable record of booms and busts, the sector's contribution to energy supply has declined somewhat recently. But coal remains the only abundant domestically produced energy resource and represents the primary fuel used by a large segment of industry.

In the eastern half of the United States, coal was formed approximately three hundred million years ago; in the West much later. The energy content of eastern bituminous coal averages 22 million Btu per ton, the equivalent of 22,000 cubic feet of natural gas, 159 gallons of distillate fuel oil, or one cord of seasoned hardwood. Eastern coal is the best coal in the nation. More than one-half of total U.S. output comes form the Appalachian Coal Basin, an area that has been mined for more than two hundred years.

The Appalachian coal industry is currently in the midst of a no-growth period and faces an uncertain future. Considerable excess mining capacity has depressed prices, and falling world prices of coal substitutes offer little prospect for a return to the coal boom years of the 1970s.

This book explores the general economic character of the Appalachian coal industry. It is set in a traditional framework of demand and supply analysis and examines the impact on the industry of the important economic variables. No particular viewpoint is supported in these pages, and no cause is represented. As is often the case, sufficient and reliable data are unavailable, and thus it has been necessary to derive conclusions from nontraditional sources. This has been done carefully and judiciously. The product is, I hope, a comprehensive review of the economics of the Appalachian coal industry and of the principal issues that face it.

This book is intended for a broad audience: researchers in the field of energy, policy makers at public or private levels, coal producers and energy suppliers, electric utility and other users of coal, students, and prospective investors in energy-producing firms.

The author is indebted for financial support to George Evans, Secretary of the Kentucky Energy Cabinet, to Lyle Sendlein, director of the University of Kentucky's Institute for Mining and Minerals Research, and to Lee Brecher, former acting director of the institute. Without their encouragement this effort would have been stillborn. Many others have contributed valuable assistance as well, in particular Janet Cabaniss and the University of Kentucky Department of Economics' indefatigable staff assistants Janice Davies and Kay Nester. All errors of commission or omission, however, are mine alone.

COAL IN
APPALACHIA

I

COAL: AN OVERVIEW

Coal is a combustible carbonaceous rock of diverse constituent elements and of variable chemical compositions. It is believed that coal was formed during the Carboniferous period 300 million years ago. As plants, trees, and other organic materials died, they did not decay completely but began to accumulate through normal sedimentary processes. They formed distinct strata, today's coal seams. It is curious that a relatively narrow coal stratum just a few feet thick can and often does extend over hundreds of square miles.

Scientists suspect that natural burial of organic materials with other organic materials occurred rapidly and that differences in types of coal are explained mainly by differences in the composite organic matter and in environmental conditions during burial. As the covered organic material began to fossilize, it built up carbon and turned into peat, then into lignite, and then into coal. Because lignite (a coal intermediate in grade between peat and bituminous coal) has never been discovered beneath coal, and peat never beneath lignite, scientists are confident about the accuracy of the progression. As lignite was subjected to heat and pressure over time, it evolved eventually into bituminous and anthracite coal. In the process it lost volatile matter but acquired carbon and thereby rose in rank (rank refers to the carbon content of coal; the higher the carbon, the higher the rank).

Coal, like any other fuel, is in demand because it contains carbon, which has heating value. This heating ability is expressed in terms of British thermal units (Btu).[1] The amount of heat produced by the combustion of coal is also expressed in Btu. The volatility of

Figure 1. Major Coalfields of the United States

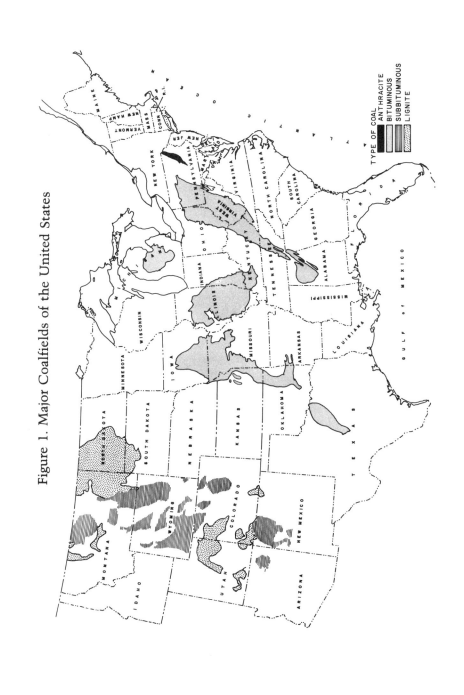

coal is the portion of the organic matter that is driven off in the form of gas. There are coals that contain high, medium, and low proportions of volatile matter; they are referred to as high-, medium-, and low-volatile coals. Each satisfies a somewhat different heating need. The heating value of fuels increases progressively with rank. Peat has the lowest heating value, followed by lignite coal, subbituminous and bituminous coals, and finally subanthracite coals.

Coal deposits exist everywhere in the world, but the largest quantities have been discovered in the Northern Hemisphere. It is estimated that one-fifth to one-sixth of the world's coal deposits are located in the United States and that they occur in thirty-seven states. Figure 1 identifies the major coal-producing regions by type of coal produced and by name of region as used in this book. Historically, the majority of coal tonnage mined in the United States has been extracted east of the Mississippi River, but this has changed, and now nearly one-fourth of all coal mined comes from west of the Mississippi. In addition to the United States, Russia and China also have large coal deposits. Together these three nations produce more than 50% of world output and own two-thirds of the world's reserves.

Sulfur dioxide emissions, mostly from electric utilities, link coal to the single most important source of environmental pollution. While the emissions themselves are problematic, a related problem is that when coal is burned, sulfur dioxide ejected into the atmosphere combines with water to form sulfuric acid; this combination can produce what is known as acid rain.

Lignites, subbituminous, and anthracite coals have the lowest concentrations of sulfur; some bituminous coals also are low—largely those located in the Eastern Interior Coal Basin—but most have relatively more. Some of the sulfur occurs in minerals that are attached to the carbonaceous materials and are inorganic. Methods of washing newly mined coal can be highly efficient in removing this pyritic sulfur. Such technology is well known and was originally developed in order to reduce the transport costs of energy. Obviously, the purer the coal transported (the fewer its impurities and the higher its quality), the lower the transport costs per Btu. Other sulfur is part of the carbonaceous material itself and is organic. Although the technology for removing organic sulfur prior to combustion does not exist, removal is possible after combustion, for example, through stack gas cleanup procedures commonly called scrubbing. In figure 2, the relationship between sulfur and carbon content is shown diagrammatically.

Figure 2. Relationship of Sulfur Content to Carbon Content
in Coals of Varying Rank

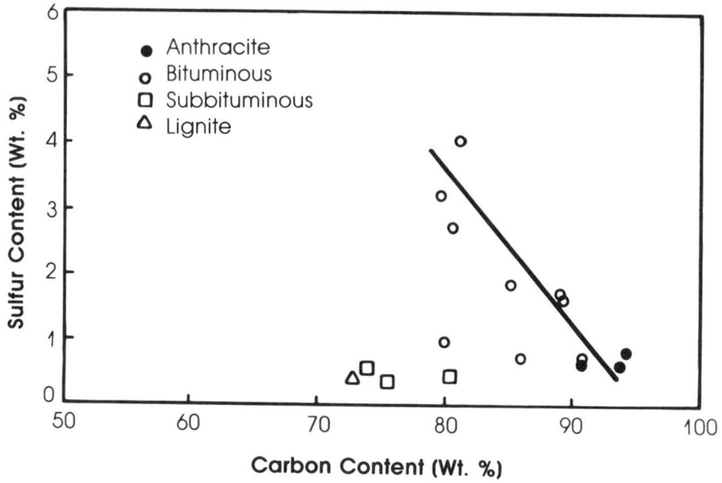

Source: *Steam,* New York: Babcock and Wilcox Company, 1972

WORLD COAL RESOURCES AND RESERVES

The world is generously endowed with coal. The reserves are extensive and widely dispersed. They also vary greatly with regard to their ability to produce heat and to the amounts of sulfur, ash, and volatile matter they contain.

Data on international geological coal resources and technically and economically recoverable reserves are extremely difficult to collect.[2] Recent increases in the role of coal in international commerce and reevaluations of it as a source of future energy have led many countries to prepare and revise estimates of their coal reserves. Table 1 ranks nations by size of their technically and economically recoverable reserves and also identifies their geological resources. (Tables begin on page 145.)

Although these estimates ought to be used with care, they do reveal some pertinent facts. For example, it is clear that even though the Soviet Union has almost twice as much coal as the United States, much of it is technically and economically unrecoverable. It would seem from table 1 that the United States possesses by far the largest accessible reserve in the world. This leads to the conclusion that the United States will most likely remain the dominant coal-exporting nation well into the future. Its coal seams, even those that are deep mined, lie relatively close to the surface, are flat, and are thus less expensive to extract. Unfortunately, this cost advantage is in many cases offset by long distances required for coal delivery and by the use of inferior mining techniques. Such factors have enabled Australia and South Africa to offer their coal in Europe at prices which are often 20%–25% below American prices. The centrally planned regime of Poland also can undercut all free-market prices because profit maximization is not one of its principal goals. At this writing, Poland's goal to earn convertible exchange to meet its international debt commitments is overriding all economic considerations. Consequently, American, Australian, and South African coal sellers are finding it increasingly difficult to compete in European markets.

The data in table 1 also highlight the point that both South Africa and Poland are not very generously endowed with geological resources. From a geological standpoint, therefore, the implication is that as recoverable reserves are depleted, Polish and South African coal will become more expensive to mine. This may also become true of Canadian coal and to a lesser degree of Australian coal. Consequently, in the long run, if world demand for coal remains strong, the U.S. position in world markets may improve over those of its competitors.

Table 2 illustrates the distribution of recoverable bituminous and subbituminous coals among nations. In general, the former are preferred over the latter, particularly if they are clean (sulfur-free). Bituminous coals have greater heating values per pound and hence are less expensive both to burn and to transport.

It is evident from table 2 that more than three-fourths of recoverable world reserves are bituminous. The United States and China each have about one-fifth of these reserves, the Soviet Union somewhat less. It is also clear that none of these nations needs be concerned about future solid fossil fuel reserves. More important would be the effect of mining these reserves upon environmental stability.

WORLD PRODUCTION OF COAL

At least forty-one nations produced coal in 1984, in quantities varying from less than 1 million tons to over 890 million tons. Table 3 ranks the major coal-producing nations in accordance with the 1984 quantity of output. These nations can be grouped into at least three categories in terms of short tons (a short ton is 2,000 pounds) produced in 1984:

1 > 600 million	2 200-300 million	3 100-200 million
United States	East Germany	Australia
Soviet Union	West Germany	South Africa
China	Poland	India
		United Kingdom
		Czechoslovakia

The gap between the annual production of the three largest producers—the United States, the Soviet Union, and China—and category 2 is rather wide. Nearly as pronounced is the gap between the nations making up category 3 and the next largest producer, Yugoslavia.

Production statistics for the past five years underscore the remarkable stability of coal production in centrally planned economies. Only in 1981 was there a significant drop in mining activity in such countries, and then only in Russia and Poland. In Poland, political unrest caused the 14% drop in output. The 2% decline in Russia is generally attributable to poor planning and inefficient mining.

The market-oriented economies of the United States, Australia, South Africa, and India have expanded production significantly in the past five years, principally because rising oil prices have made it economically logical to substitute the use of coal in production processes for the use of oil (a substitution effect). These nations own ample coal reserves, and production takes place under favorable technological, geological, and economic conditions. In contrast with the centrally planned economies and those of England and West Germany, these economies can expand easily and (usually) quickly, and any future escalation in demand will most likely be satisfied by production from these nations.

U.S. ENERGY CONSUMPTION

In the post-World War II period, energy consumption has grown rapidly as a result of growth in the U.S. economy. The rapid pace of industrialization makes it unsurprising that energy consumption grew considerably faster than population.

Table 4 shows gross energy consumption in the United States for the period 1950–1984 per dollar of real gross national product

(GNP) produced and per resident (see columns 5 and 6, respectively). The data in column 5 show that up to about 1970, energy consumption per dollar of GNP was nearly the same as twenty years earlier. Relatively speaking, energy was an inexpensive factor input, and consequently there was little incentive to conserve it. Capital equipment and durable goods were being designed with an eye toward conserving labor inputs, with the amount of other energy consumed of little consequence. All this changed in late 1973 with the imposition of the OPEC (Organization of Petroleum Exporting Countries) oil embargo and the subsequent rise in petroleum prices, but it took several years for the impact to be reflected markedly in energy-per-dollar-of-GNP statistics. In fact, we are just now seeing the full impact of the 1973–1974 and 1979 increases in oil prices.

The reasons for this time lag are fairly straightforward. In the short run, it simply is not feasible to use substitute fuels when the price of one fuel increases. The cost of abandoning equipment merely because the price of oil had risen was too great. Also, in the short run it is not possible to improve the efficiency of existing capital stock in response to higher fuel input costs. Machines and equipment are generally designed and built to accommodate a given technology of energy consumption and cannot be altered easily.

Immediately following the oil boycott, efforts were made to substitute natural gas for oil wherever this close substitute could be used. The economic incentive existed because prices of delivered residual oil increased in 1974 by 143% while natural gas prices rose by only 42% (see table 5). In 1973 the price ratio of residual oil to natural gas was, on average, 2.3; by 1974 it had risen to 3.9, a 70% increase. Natural gas prices were prevented from rising more rapidly by selective statutory price ceilings, which continue in force in 1985. In part it was these price ceilings that were responsible for the 1977 natural gas shortages.

In 1974 the price of coal leaped upward, spurred by utility companies' concern that they might be asked to rely entirely on coal-fired furnaces in the future. Even though coal prices rose by 75% in 1974, they remained much below oil's new prices.

In 1981, following the 1979 second round of oil price boosts, the oil-to-gas and oil-to-coal price ratios were 1.9 and 3.5, respectively. This relationship illustrates that coal is by far the less expensive fuel input if the costs of environmental air emissions are disregarded.

Quite another story is revealed by per capita statistics on energy consumption. Fueled by low gasoline prices, U.S. demand for oil expanded rapidly in the post-World War period, reaching new heights in 1973. The oil embargo and price increases dampened this demand sharply, but consumers made a relatively speedy adjustment. Expenditures on other goods and services were scaled back, and per capita energy consumption once again returned to its 1973 level. Not until 1980, mostly in response to the 1979 doubling of oil prices, did energy consumption begin to decline once again. The new price of thirty-four dollars per barrel had a debilitating impact on economic growth worldwide. As discretionary spending declined and industrial, residential, and automobile fuel conservation measures began to take hold, energy consumption fell sharply. By 1981 per capita energy consumption had declined to 1970 levels, having experienced a 10% drop in just two years. This trend is likely to continue, particularly because of uncertain and unstable economic conditions worldwide. Most forecasting models that have predicted future energy consumption trends and prices, even the most recent ones, have significantly overestimated energy consumption levels. This is, perhaps, not surprising, because econometric forecasting depends on historical data. Even when these data are adjusted to reflect expectations of continued high fuel prices, the historic bias is unable to accommodate adequately the effects of the occurrence of shocks and disturbances that were assigned a low probability by the models. Thus, these econometric forecasts must be supplemented with judgment.

Another aspect of analyzing energy consumption is by type of fuel. Table 6 shows U.S. gross consumption of coal, petroleum, and natural gas from 1950 to 1984. It also depicts nuclear energy and hydropower consumption for the period.

The supremacy of petroleum and natural gas as energy fuels between 1950 and 1973 appears unchallenged. By 1973, petroleum consumption had increased to 2.5 times its 1950 level and natural gas consumption had increased to more than 3.5 times that level. Coal consumption, on the other hand, increased by only 3% for the same period. Oil and natural gas were the low-priced fuels of that period, and it is not surprising that the growth in energy demand was satisfied by them. From 1973 to 1978, growth in energy consumption was modest as the economy struggled to absorb and adjust to the higher prices of fuels. Only nuclear power consumption grew significantly during this period. This form of energy clearly is less sensitive to price-induced substitution and income effects than are the fossil fuels. Commitments to nuclear power are generally long-term and uninterruptible.

In the latter part of the period shown in table 6, the doubling of oil prices in 1979 took its toll on petroleum consumption. Beginning a sharp decline in 1980, by 1983 petroleum consumption was 21% lower than its high in 1978. The recession that ensued depressed the consumption not only of petroleum but of natural gas and hydropower as well. Only the consumption of coal grew, due exclusively to its increased use by public utilities in the generation of electricity. By 1980, half of the electricity produced in the United States was coal based. From 1979 to 1984, electric utilities increased their consumption of coal by 26% as interfuel substitution, particularly that of coal for oil, became more pronounced.

DOMESTIC DEMAND FOR COAL

The demand for coal, like the demand for all factor inputs, is derived from the demand for the final products in whose production it and other inputs participate. The demand for coal as an end-use commodity is small indeed, confined as it is to that portion of the market in which coal is burned directly by consumers for home heating.

By far the largest source of demand for coal originates with the electric-power-producing companies—the electric utilities. While all other sources of demand for coal experienced declines in the early 1980s, the electric utility demand kept growing, albeit at a reduced rate. Table 7 illustrates the origin of the demand for coal by economic sector for the period 1970–1984. From 62% of total U.S. coal consumption in 1970, electric utilities' consumption grew to 83.8% in 1984. Most of the very weak increase in the production of electric energy in 1981 was accomplished by increased use of coal-fired furnaces.[3] In short, more than four-fifths of the coal consumed in the United States is burned at electric utility companies. It is obvious that the welfare of the electric utilities dictates the welfare of the U.S. coal industry.

The remaining important sectors are the export and "other industrial, including transport" sectors. Export demand for U.S. coal is considered in a separate chapter of this book. The "other industrial" category is one that saw its demand for coal decline sharply up to 1973, mostly because oil and natural gas were being substituted increasingly for coal. After 1974 this decline was arrested as some industries substituted coal-fired furnaces for oil- and natural-gas-fired ones; the 1977 natural gas shortage and the doubling of oil prices in 1979 were key factors here. Even so, in 1981, for example, coal flowing to the "other industrial" category did not exceed 9% of total coal consumption.

Unlike the aforementioned category, the demand for coking coal has declined steadily in the past eleven years. From 1970 to 1984 this decline equaled 51% of 1970 production. In 1984 the market represented only 5.5% of total U.S. consumption. Compared with the electric utility market, the coking market is relatively small and declining.

U.S. COAL RESERVES

Coal is the most abundant U.S. fossil fuel resource, and its reserves are widely distributed across the nation. Unfortunately, it is also the most difficult fuel to mine, transport, and handle. In quadrillions of Btu, the following ranking can be constructed of the amount of energy stored in domestic fuel reserves. While this ranking reveals the estimated raw energy embedded in the respective fuels, it says nothing about the technical or, more important, the economic feasibility of harnessing it. It does underscore the overwhelming relative abundance of coal.

	Quadrillions of Btu	Percent of Total
Coal	4,800	85
Oil shale	430	8
Natural gas	200	4
Petroleum	160	3
Heavy oil	25	–
Tar sands	15	–
Total	5,630	100.0

Source: Electric Power Research Institute, EPRI Journal vol. 6, no. 6 (July/August 1981).

The most meaningful measure of coal resources in the United States is the demonstrated reserve base. This measure includes both geologically indicated and actually measured seams of bituminous coals at least 28 inches thick and of subbituminous coals at least 60 inches thick that lie at depths no greater than 1,000 feet. It also includes selected thinner seams that can currently be mined economically.

The Energy Information Administration of the U.S. Department of Energy estimated that on January 1, 1983, the demonstrated reserve base included 490 billion tons of coal, 20 billion tons more than the 1979 estimate. The increase reflects readjustments in reserve estimates.

A taxonomy of the coal reserve base is presented in table 8. It shows the amount of unmined U.S. coal, and whether it is minable with underground or surface methods. A third of all coal is surface minable, and nearly three-fourths of such coal lies west of the Mississippi River. More than half of total reserve tonnage requires underground mining, lies mostly to the east of the river, and is of bituminous rank. Subbituminous coal, 37% of the total coal in reserves, is found entirely west of the river; 9% is lignite, which exists also almost entirely in the West.

Not all of the coal reserves included in the demonstrated reserve base are recoverable at this time. In fact, the recovery rate, the percentage of coal that can be recovered from the reserve base, ranges from less than 40% in some underground mining to greater than 90% for some surface mining. In general, however, standard recovery rates of 50% for underground and 80% for surface mining are considered reasonable average estimates.

Coal reserves by state and by type of mine are shown in table 9. Surprising is the fact that two of the three states with the largest underground reserves are located in the West: Montana and Wyoming. These two states alone hold more than a third of total U.S. underground reserves. Even so, in 1980 Montana mined no coal underground, and Wyoming mined a mere 1.6 million tons, only 2% of its total production. At present both states surface mine because of the low Btu content of their coal and the enormous seam sizes, factors that make it far more economical to surface mine than to deep mine. This situation is likely to persist because, as table 9 shows, between them the two states contain nearly two-fifths of all U.S. surface-minable reserves. At 1980 production rates, if other factors remain constant, these reserves alone would not be exhausted for hundreds of years. It is therefore not surprising that mining activity in these two states is confined to the surface of the land. The cost of deep mining would exceed considerably the value of the product in the marketplace.

The coal reserve base in Appalachia is particularly important because it is the source of most coal consumed in the United States today. The demonstrated reserve base for the region is shown by state and mining method in table 10. The largest total reserves are located in West Virginia and Pennsylvania and consist mostly of underground coal. Together these two states account for two-thirds of the Appalachian Coal Basin total. Ohio and Eastern Kentucky contain the next two largest underground reserves.

Surface-minable coal deposits are more evenly distributed than underground reserves. Ohio, West Virginia, Eastern Kentucky, and Alabama hold the largest amounts of these. Most of the present and probable future surface-mining activity in Appalachia is likely to take place in these states.

Somewhat more than one-fourth of the underground and one-eighth of the surface reserves in the United States are located in Appalachia. The coal located in Appalachia therefore represents a relatively small portion of the national endowment, but it is the best available in terms of heating value and purity. It is lower in sulfur content, higher in Btu, and lower in ash than competing coals. The ranges of the ash, sulfur, and Btu content of the coal located in the four principal coal-producing states in Appalachia are shown in table 11. Some of the best coal is found in Eastern Kentucky, the least desirable is in Ohio, with West Virginia and Pennsylvania coal spanning the remainder of a very broad spectrum. If ambient air standards are tightened in the eastern United States, it is quite likely that coal low in sulfur and ash will face increased demand; spurred by escalating prices, production of this type of coal will rise.

DOMESTIC SUPPLY OF COAL

Following World War II there developed a large-scale switch from coal to the new glamour fuels: oil, natural gas, and, somewhat later, nuclear power. Railroads switched to diesel fuel in the United States and abroad, and other users in industry and manufacturing followed suit. Oil and natural gas were more convenient fuels to use, and the relative levels of prices made the shift economically attractive. In the 1960s the coal industry was in a production trough, but by the end of the decade it had begun to recover from the blow of a 36% decline in output (see table 12). By 1970, annual production had recovered to a level of 603 million tons. In subsequent years, in a response to the 1973 OPEC oil shock and other factors, interest in coal as a future energy source resurged. For the first four years of the 1970s, annual coal production remained relatively constant, but beginning in 1975 it grew rapidly to 830 million tons in 1980. The next year saw a modest decline due principally to the effects of a weakening U.S. economy. In fact, what kept coal production in 1981 at relatively high levels was unexpectedly strong export demand. The upsurge in 1981 exports, an increase of 23% over 1980, was largely because of foreign purchases of U.S. steam coal in the spot market. Political unrest in Poland and widespread strikes in Australia had led to serious disruptions in the delivery of coal from these countries. Concerned European buyers sought out U.S. coal producers to make up their deficiencies. The coal export boom began to gather force in late 1980 and lasted through 1981 and part of 1982. By the end of that year, however, it had pretty much run its course, so that the last few months recorded weak shipments abroad.

In addition to overall changes in coal production statistics, some very pronounced structural changes have occurred in the past twenty years in regional outputs. Table 13 shows the origin of coal output by state and region. The most notable change took place in the western coal region. In just twenty years Montana and Wyoming developed into major coal-producing states. In 1960 these two states produced a mere 2.3 million tons, 0.5% of national output. By 1984, however, these two were producing 161.2 million tons, 18% of national production. The other states in the region also evolved into important contributors to the coal market, so that in 1984 western coal region producers were mining considerably more coal than were the Eastern Interior Coal Basin operators.

The explanation for the phenomenal growth in western coal output lies in two factors: (1) the low costs of surface mining coal in the West due to very large seams and small overburdens, and (2) the low sulfur content of these coals. Despite the sizable costs added on by the distance of the end user from western mines, mining costs are sufficiently low in the West to enable such coal to compete successfully in markets previously considered the exclusive domain of coal from the interior basin. In short, western coal suppliers have made increasing inroads into markets historically serviced by the high-sulfur coal produced in Illinois, Indiana, and Western Kentucky. As table 13 shows, while coal production east of the Mississippi River between 1980 and 1984 remained nearly unchanged, western output increased by 22%.

II
PRINCIPLES OF ENERGY RESOURCE ALLOCATION

Before examining the economics and the structure of the Appalachian coal industry, a review is in order of the framework within which this examination will be conducted. The approach pursued here is to analyze the industry in terms of governing demand and supply relationships. Also considered are legislative, sociological, and historical influences insofar as they affect the demand for and supply of coal. No new modeling is undertaken for this book, but relationships discovered in past modeling efforts are discussed. Their value lies in their ability to enhance our understanding of how the industry has functioned in the past, what some of its problems are today, and where the years ahead may take it. Of importance are, among other factors, the influence of the world oil market, the impact of pending environmental legislation, the future of the domestic and world economy, and technological progress. This is an eclectic approach that is strongly anchored in the fundamental principles that guide economic analysis.

ECONOMICS OF ENERGY RESOURCE DEMAND

All too often coal-industry specialists fail to consider adequately, and in time, that industry health and survival are affected most strongly by the demand for its product. Demand is influenced by the prices and availability of fuel substitutes, by world politics and current events, by the economic well-being of literally dozens of countries, by government regulations, and by many other factors.

Like the demand for other natural resources, the demand for coal is derived from the demand for the final commodity or service to which it contributes. While in the past consumers used coal directly for heating homes, this portion of the demand for coal has pretty much passed into history. Today more than 80% of the coal consumed in the United States is used to generate electric power; less than 1% finds its way into residential markets.

The relationships between the demand for coal and the various factors that influence it are frequently summarized in mathematical shorthand. A functional representation describes the quantity demanded (D_c) at any point in time as dependent on p_c, the price of coal; $p_1, p_2, \ldots, p_n$, the prices of fuel substitutes; $q_1, q_2, \ldots, q_n$, the quantities of electric power, iron, steel, and other products produced with the aid of coal; and $k_c, k_1, k_2, \ldots, k_n$, the qualitative properties of coal and of its substitutes (for example, carbon, sulfur, and ash content) that influence demand decisions on coal:

$$D_c = f(p_c; p_1, p_2, \ldots, p_n; q_1, q_2, \ldots, q_n; k_c, k_1, k_2, \ldots, k_n)$$

The expected influence of the prices of fuel substitutes for coal is inverse: as long-run prices rise more coal will be demanded, and as they fall less coal will be demanded.

The key variable that determines the quantity demanded of any commodity in the long run is its price. Coal is no exception, and the quantity of it demanded is an inverse function of its price.[1] This means that if the influence of other factors is set aside, the lower the price of coal the more of it will be demanded. Conversely, the higher its price, the less will be demanded. The exact price-quantity relationship—that is, the sensitivity of the demand for coal to changes in its price—is captured in the principle of price elasticity of demand and can generally be measured empirically.[2]

In general, price elasticities vary greatly across commodities and with respect to time. For example, when the price of gasoline rises abruptly, consumers reduce their consumption only slightly because few, if any, substitutes are available to them. When the price of fish increases at market, on the other hand, consumption declines substantially; substitutes in the form of beef, pork, or fowl abound. Studies indicate that the elasticity coefficient for gasoline is -0.2; for fish, -2.2.[3] This means that a 1% increase in the price of each would result in but a 0.2% decline in the quantity demanded of gasoline but in a 2.2% decline in the quantity demanded of fish. Revenues collected by businesses selling the commodities would increase in the former, decline in the latter. The elasticity concept also applies to the quantities demanded of all energy resources.

Implicit in this concept is the principle of substitution. If, as the price of an energy resource rises, little or no reduction occurs in its quantity demanded, the substitution of other commodities for the now-higher-priced energy resource is not occurring.[4] Instead, consumers are paying the higher energy costs. They are able to do so, out of a given budget (income), only by reducing their expenditures elsewhere. In essence this is what happened after the 1974 oil price increases and once again, to a somewhat lesser extent, in 1979. Consumers absorbed the higher costs of oil, as if a foreign tax had been levied on the commodity, by cutting back expenditures on other goods. A general decline in economic activity ensued, while the share of energy expenditures in gross national product (GNP) rose consistent with the increase in prices. A substantial international transfer of income from the oil-consuming to the oil-producing nations began to take place.

If, following the 1974 and 1979 oil price increases, consumers had reduced their consumption of oil and other energy resources significantly, and if therefore the share of spending on energy in spending for total output of the nation had declined more than proportionately to the price increase, this would have been a desirable situation indeed, at least from the standpoint of the United States. Such a situation would have signified that the demand for energy resources was highly price sensitive and that no income transfers from oil-consuming to oil-producing nations were likely to occur.[5] What consumers did not spend on energy they would have spent on other goods. Under certain assumptions total output of the economy actually could have increased. This would have been an ideal solution to hikes in the price of oil. Unfortunately, this is clearly not what happened. Consumers largely maintained their energy expenditures, and total output of all goods declined. The evidence shows that the demand for energy in the short run is inelastic, not elastic. In other words, price increases in energy resources do not evoke a substantial reduction in the demand for them in the short run. There is little that can be substituted for energy resources in the near term.[6] In the long run, however, a different picture emerges.

Given sufficient time, increases in the price of energy resources can bring significant changes in the quantity demanded. As the prices of energy resources change in relation to those of non-energy goods, consumers of energy have sufficient time to adjust to altered price ratios. For example, in the long run, manufacturers may begin to substitute labor resources in production for some capital resources because the latter consume far more energy than the former. If such a substitution is technically feasible, particularly at the margin of production, the result may be a lower level of demand for energy under the new pricing structure and higher demand for labor. But the productivity of labor is bound to decline in the long run as labor has fewer technically useful capital goods to work with and production becomes more labor intensive.

Over time, rising energy prices also affect technological change. New equipment, machinery, and production techniques may be developed featuring energy-saving devices. When such devices are used in production,[7] the overall quantity of energy resources used and needed will be altered in perpetuity, and the goal of energy conservation will have become an institutionalized part of technical change. To a very large extent this has happened in the past ten years.

The decline in the real price of energy resources between the middle 1960s and the early 1970s led to their increased use. This fact is reflected in part in the reversal of the rising postwar trend of the ratio of energy consumed per dollar of GNP. The oil boycott put an abrupt end to this trend, and the energy-to-GNP ratio began a decline. Between 1973 and 1981 the ratio declined 24%, a trend anchored in the irrevocable change in energy consumption practices (see table 4).

Unlike short-run happenings, the long-run possibility of interfuel competition and substitution makes the demand for energy resource sensitive to changes in its price. If changes in the relative price of an energy resource are sufficiently great and are perceived as permanent, the incentive to retrofit existing equipment, such as boilers to produce steam, becomes real. The costs of conversion are extremely high, however, particularly from liquid to solid fuels. Decisions to do so are not made quickly, especially in a world of uncertain long-term trends.

It is instructive to examine the results obtained from numerous studies of the own-price elasticities of demand (that is, the effect of a change in the price of good A on the quantity demanded of good A) for oil, natural gas, and coal in the long run. Table 14 summarizes these results by demand sector. Each coefficient shows the percentage change in the quantity of fuel demanded in the long run due to a 1% change in its unit price. For example, according to Griffin (1977), in the electric power industry, a 1.0% increase in the price of coal would result in a 0.5% – 0.8% decrease in its quantity demanded. According to Atkinson and Halverson (1976), this decrease would lie between 0.4%-1.2%.

Differences in the coefficients listed in table 14 are due mainly to differences in estimation techniques, in data bases, and in their quality. In the residential and commercial sectors, no estimates are available for coal because in this market it has been replaced by oil and gas. But in industry and for the electric utilities, coal is still an important fuel. Because coal is less attractive to handle and burn than the substitute fuels, its price elasticity of substitution is higher than that of oil, at least in the Pindyck study. In the Baughman-Zerhoot study (1975), there appears to be little difference in the relative attractiveness of the three fuels, but coal's elasticity coefficients appear to be lowest.

In the electricity-generating industry, the elasticity coefficients for coal are the lowest among the three fossil fuels. The explanation may lie in institutional as well as economic factors. In most coal-producing states it would be politically infeasible to substitute oil or gas for coal to produce electric power, regardless of the price of coal. And there are many more coal-producing than oil- and gas-producing states in the nation. This inflexibility would lead to a low price elasticity of substitution. Most industrial states, on the other hand, have no energy resources and therefore have no similar vested interests. Their demand for the other fuels is more price elastic.

In short, the evidence from the studies listed in table 14 shows that in the long run the demand curves for oil, natural gas, and coal are moderately sensitive to changes in their prices. Income and substitution effects do occur.

DEMAND FOR U.S. COAL

By far the largest portion of the total demand for coal (85% in 1983) originates with utilities that use it as a boiler fuel for the generation of electric power. The manufacturers of coke also use coal as an indispensable raw material. In addition, minor uses of coal include its consumption in industry, in commercial and household heating, and in the railway industry. In the past, foreign buyers of U.S. coal used it almost exclusively for the making of coke, but more recently, starting especially in 1980, they began using U.S. coal also as a boiler fuel. Irrespective of the national origin of the demand, coal is used today for two basic purposes—as a fuel burned to generate steam (which uses steam coal) and as a raw material in the making of coke (using metallurgical grade or coking coal). Metallurgical coal is distinguished from steam coal by its low volatility and coking properties. Because coal with these properties is relatively scarce, coking coal commands a higher market price than steam coal—often twice the price of the latter. Producers are reluctant to sell coking coal at steam coal prices, even though it is possible to mix these two grades of coal; however, whenever metallurgical coal prices are depressed it is not uncommon for it to be sold at lower prices and blended with steam coals. Basically it is correct to differentiate between these two types of coal and to treat the market for each as independent of the other.

Since 1960 the railroad demand for coal has declined to near zero, and the retail-household demand has also become insignificant. Consequently, one can either ignore these two sources of demand or include them in the category of the industrial demand for coal.

It is wise to define an export demand category for coal because the factors that influence foreign demand are different from domestic demand factors. They include exchange rate fluctuations; competition from coal produced in Australia, South Africa, and Poland; and foreign air pollution standards.

Thus, four groups of consumers of U.S. coal can be identified. In order of size, they are (1) the electric utility industry, (2) the export market, (3) industry-railroad-retail, and (4) the coke-manufacturing industry. Of these four consumer groups, the electric utilities have recorded steady growth in the post-World War II era. In 1981 and 1982, export demand also grew rapidly, but demand stabilized in early 1983 and thereafter began to fall. The remaining two groups have steadily decreased their demand for coal in the past eleven years.

Transportation costs are a more significant factor for coal than for any other fuel and can make up 20% – 80% of the delivered price. Because these costs are so high, three basic U.S. geographic markets for coal have emerged: the eastern, the midwestern, and the western. This is not to say that western coal is excluded from eastern markets, or that Appalachian coal is not shipped to the Midwest. But it is true, for example, that in the past all of the coal burned in Kentucky was mined in the state, and that Indiana uses mostly Indiana coal. An exception is the recent penetration of western coals into eastern markets. A substantial portion of Montana's and Wyoming's production is shipped east now, and this has begun to blur traditional market boundaries and loyalties.

In a recent effort to develop an econometric forecasting model of coal prices and the distribution of production, Ali, Harvey, and Stewart developed estimates of the own- and cross-price elasticities of the demand for coal and of the final-product elasticities.[8] These estimates cover the short- and long-run time periods and the three U.S. geographic regions. An additional type of that demand, by foreign consumers for exports, was separated by region into Japan, Canada, and the remainder of the world. A summary of the results is shown in table 15, which shows, for example, that a 1% increase in the price of Appalachian coal consumed by electric utilities located in the eastern United States would result in a 0.56% reduction in quantity demanded. On the other hand, a 1% increase in the price of oil would result in only a 0.09% increase in the quantity of coal demanded by eastern electric utilities, a statistic that implies very little substitution of coal for oil indeed.

In terms of the final-product elasticity, a 1% increase in electric power production leads to a 0.88% increase in coal consumption in the eastern United States, close to a one-to-one ratio. In the Midwest a 1% increase in electric power production would lead to a 1.54% increase in the quantity of coal demanded from the Eastern Interior Coal Basin.

The results obtained for the other three demand categories should be interpreted similarly. For example, a 1% increase in the price of metallurgical coal would result in no reduction in quantity demanded in the midwestern United States. A 1% increase in the long-run output of pig iron in the western states would result in a near proportionate—0.98%—increase in the quantity of metallurgical coal demanded.

The export-sector results are more difficult to interpret. For example, it is not altogether clear why a 1% increase in electric power production in Japan should result in a 3.08% decrease in Japan's demand for U.S. coal. Further analysis would be needed to explain these results. It is possible to speculate, however, that the cost structure of electricity production in Japan is such that as the demand for electric power increases, economies of scale lead to the substitution of oil- and gas-fired furnaces for coal-fired furnaces.

In general, the study results support the expectation that the own-price elasticities for coal are negative and that the cross-price (substitution) and final-product elasticities are positive. The results also show that for most demand equations the coefficients of the time variable are significant and negative. Because the demand for coal is derived, such results imply that technological progress is steadily reducing the amount of coal required in the production of final commodities. Not surprisingly, the findings also show that the quantity of electric power and of steel produced in any one time period determines the amount of coal consumed by either industry. With only a few exceptions, oil and natural gas appear to be weak substitutes for coal consumed by the utilities and by the industry-railroad-retail sectors. An exception may be utilities located in the Midwest, where some may have a modest capability of substituting gas for coal. Finally, as expected, coal prices are a major determinant of coal consumption for the different coal consumption groups, except for coke manufacturers.

The study results also show that as a rule consumers of coal, other than coke manufacturers, take approximately two years to adjust to a disturbance that changes the desired level of consumption from the existing one. Coke manufacturers take only a little more than a year to make such adjustments, while the suppliers of coal take approximately three years to adjust.

Final-product elasticities are generally high, and, as expected, there exists no substitute for coal in the coke manufacturers' sector or in export demand.[9]

In addition to price, the properties of coal also have an influence on the demand for it. Coal users buy energy, and, other things equal, the more of it per dollar of expenditure they receive, the lower will be their unit costs of production and hence the more of it they will demand. Coal users are also concerned with the sulfur and ash content of the coal they buy. Air purity standards are not uniform across the country. In certain locations the burning of high-sulfur and high-ash coal is prohibited; in others, prohibitions are less stringent.

ECONOMICS OF ENERGY RESOURCE SUPPLY

Many factors other than demand affect the supply of energy resources. Such factors influence mostly the cost of production—interest on capital borrowed; wages paid to labor; productivity rates; health, safety, and environmental regulations; depletion; and royalty rates. Other factors are political upheavals, international alliances, and even strategic decision making. The list is very long indeed, and any one of these variables can seriously interrupt or speed up the flow of energy to markets.

If only the economic factors are considered, it is convenient to group these into four categories. First, the quantity of any energy resource supplied depends on the goal of the supplier. One would expect that this goal is profit maximization. However, this concept can be variously interpreted, and it has both short- and long-term dimensions. What may seem not to be profit-maximizing behavior in the short run may in fact be part of a strategy of profit maximization in the long run. The economic literature on optimal extraction rates and paths addresses this issue. For example, in order to maximize long-run profits, an energy producer—a coal mine operator or an oil or gas producer—must consider the future and select an extraction path that yields the highest present value for a given set of reserves. If all individual producers behave in such a rational way, then society as a whole will be maximizing the flow of revenues from its resource endowment. For example, if future prices of energy resources are expected to rise, an incentive exists to withhold supplies from markets. If energy prices are expected to fall, as seems to be the case in the middle 1980s, extraction would tend to accelerate.

Another set of factors that influence the supply of energy resources is the state of technology used in their extraction. If a technological breakthrough occurs in the production of a given energy resource, then, other things equal, unit production cost will decline and more of the energy type will be produced in the long run. Substitution of one resource for the other will have occurred.

The same principle applies to producers extracting the same resource. If one or a number of them implement a new extraction technology, the others must follow or accept shrinking sales.

A third set of factors that determine the quantity of a resource offered for sale includes its price, the prices of its substitutes, and the prices of non-energy goods. If the typical resource extractor is guided by the profit motive and if costs are held constant, a higher price will attract more of the resource to market. In part, the amount of one energy resource offered for sale also depends on the prices of the related resources. If, for example, relative prices change and profits in one energy industry rise, it is conceivable that investment capital will flow out of the other resource industries and into the more profitable one. Production of the profitable resource will expand, and downward pressures on its price will ensue.

Finally, the quantity of energy resources supplied also depends on the costs of producing them. Underlying the total unit costs are the costs of factor inputs. If any factor price increases while all else remains the same, producers will have less of an incentive to produce, because profitability will have declined. In a purely competitive market, producers would be unable to pass on the higher factor costs out of fear of losing customers.

Examples of increased factor costs abound: higher wage rates resulting from new union contracts, higher royalty or capital costs, higher licensing costs, rising materials costs. Any of these increases, among others, can materially alter the per unit cost structure of energy resources and precipitate a long-run change in use.

A general functional form of expressing the relationship between the quantity of an energy resource such as coal supplied and the factors that influence this supply is written as

$$S_c = f(G; T; p_c; p_1, p_2, \ldots, p_n; r_1, r_2, \ldots, r_m)$$

where G represents the goals of the coal producer; T, the state of mining technology used by the representative firm; and p_c, the price per unit of coal produced—either in terms of dollars per ton or per million Btu. Variables $p_1, p_2, \ldots, p_n$ can be viewed as the prices of all commodities produced, including coal substitutes in the long run, and $r_1, r_2, \ldots, r_m$ are the costs of the inputs, such as labor, capital, royalties, and others. The functional expression shows that the amount of coal that firms are willing to supply at any time is a function of the variables described in the brackets.

The most influential variable that affects quantity supplied is price. The higher the price, the more will be produced and supplied; a positive relationship holds for price and quantity.[10] The exact nature of this relationship, that is, how sensitive a change in the quantity supplied is to a change in price, is captured by the principle of the price elasticity of supply. This concept, similar to the already reviewed concept of the price elasticity of demand, is measurable empirically. It conveys the notion that if, other things equal, quantity supplied remains relatively unchanged in the face of a change in price, supply is considered price inelastic. On the other hand, if quantity supplied changes significantly in response to a price change, supply is said to be elastic, or sensitive to market price changes.

In contrast to the number of studies of price elasticities of the demand for energy resources, few have been published regarding the price elasticities of supply. The explanation probably lies in the problem of unavailable data—many of the relevant statistics are considered proprietary. The few data sources that are available are often of questionable accuracy.

The degree of supply elasticity with respect to price found in the market at any one point depends in very large measure on the time frame of reference. The longer the time span under consideration, the more sensitive is supply to change in price. The long run allows ample time to adjust production schedules, expand plant capacity, train and employ new labor, and, in short, take the necessary actions that would allow a producer to expand output. In the short run insufficient time exists to do these things. Consequently, a change in price would not elicit much of a short-run change in quantity supplied.

The concept of the long run varies from industry to industry. It is generally defined as the time needed to expand existing production facilities, build entirely new ones, or both. If it is assumed that it takes from six to eight years to open an underground coal-mining operation, then the long run for the underground-mining industry is at least six to eight years.

Another set of factors that determines the length of a particular time concept is the availability of labor and material inputs. If these are readily available, or are obtainable with only a minimum delay, they would not be considered binding constraints to the expansion of output. Also assumed is a reasonably stable factor input price.

A summary of estimated price elasticities of supply for the fossil fuels is presented in table 16. As expected, the elasticity coefficients show that supply is much less price elastic in the short run than in the long. Given sufficient time, firms will expand output in response to rising prices and will reduce it when prices decline.[11]

When compared with the elasticities of most other industries, the long-run elasticity coefficients are lower for the extractive industries. It is an observable and irreversible axiom that the most attractive mineral reserves are mined first. Consequently, as mining progresses, less accessible and deeper reserves (or narrower coal seams) are pressed into production. The time required to increase output in response to a price rise lengthens, and supply becomes less elastic. Price must be high enough to cover the higher production costs; otherwise the new energy resources would not flow to markets. The owners of the older reserves that have been exploited for some time stand to reap economic rents.

SUPPLY OF U.S. COAL

To a very large degree the willingness and ability of an industry to supply a steady flow of products to market is determined by its structure. A highly concentrated industry, in which one or only a very few companies control output, is less likely to respond quickly and sufficiently to changes in consumer demand than an industry that has many independent producers. The coal industry falls into the latter category. Output is unconcentrated, and of the energy resource industries it is by far the most competitive. In general economists hold that if the four largest firms in an industry produce less than 50% of total output, concentration and potentially noncompetitive behavior are unlikely to exist. In the coal industry the four largest firms did not produce even 20% of total 1981 output, and the fifteen largest firms produced 318 million tons, only 39.68% of total. On the basis of the market share of the largest producers, therefore, it is clear that the U.S. coal industry is unconcentrated. Table 17 lists in declining order of size the fifteen coal-mining firms that produced 40% of U.S. output. Of these, the Peabody and Consolidation groups, traditionally the two largest coal producers in the nation, continue to occupy the first and second positions, although their share in total output has declined markedly. In 1973 these two companies produced 22% of U.S. output, and in 1983 only 12%, a decline of 50%. During this time period other companies, in particular the AMAX and Texas Utilities groups, have increased their output shares considerably.

The assertion that coal production is not concentrated is also supported by an examination of the share in output held by the four, eight, and ten largest producers. The four largest produce 21%, the eight largest 30%, and the ten largest one-third of national output. Consequently, collusion and anticompetitive behavior are unlikely in the coal industry.[12]

Even if the U.S. coal market is divided geographically into more narrowly defined submarkets, concentration cannot be found. Professor R.E. Shrieves divided the U.S. coal market into fifteen multiple-state and single-state markets and discovered that seller concentrations were low in all of them. These findings led him to conclude that the sharp increases in coal prices during the middle 1970s are not attributable to market power of coal producers.[13]

Another useful way to examine the potential of producer concentration in the coal industry is by ownership of coal reserves. A survey of the demonstrated reserve holdings of the 100 largest owners conducted by the Federal Trade Commission revealed that ownership is quite widely dispersed among companies.[14]

A substantial portion of U.S. coal reserves is held by oil companies. Since 1966 virtually every major oil company has entered the coal industry, although not all have remained in it. Occidental Petroleum, Ashland Oil, Standard Oil of Ohio, Exxon, and Sun Oil Company have joined Gulf Oil and Continental Oil Company as coal producers. In the past three years Occidental Petroleum has begun to reduce its holdings and Ashland Oil has practically eliminated its coal-industry activities.

The static theories of competition emphasize that the most important element that fosters concentration in industries is the existence of barriers to the entry of new firms.[15] Although there is some disagreement among scholars as to the exact definition of entry into an industry, most agree that no monopoly can endure unless new firms are prevented from entering. If new firms can enter an industry freely, they will bid away business from existing firms by cutting prices. The inevitable consequence is profits that are lower than monopoly profits. In short, in order to perpetuate concentration and monopoly profits, existing producers must make entry into their industry costly for new entrants.

One of the key obstacles to entry into underground mining is the size of the required investment. To construct a typical underground mine having a capacity to produce 1 million tons per year requires $30 million to $40 million. The investment is indivisible and can be expected to be unprofitable for at least seven years after opening.

For a surface mine, initial capital investment is much smaller: one can operate a surface mine with just a leased bulldozer, front loader, and truck. A large portion of such equipment used in surface mining is also used in the heavy construction industry for the building of roads, bridges, and commercial facilities. Consequently, this equipment is substitutable between these two sets of activities. When construction activity slows, the machinery becomes available for surface mining. And because it is mobile, it can be easily transferred from one region or industry to another. Of course variations in the initial capital requirements can be significant, depending on geological structures and formations, soil stability and composition, seam width, and depth of the overburden.

Both the passage of the Federal Coal Mine Health and Safety Act of 1969 and the Surface Mining Control and Reclamation Act of 1977 have raised substantially the entry costs into mining. In underground mining, prospective operators now have to submit for approval detailed plans for safety equipment, installations, ventilation systems, and miner training and evacuation procedures, as well as other information. The days when a family of cousins and kin could operate a small drift mine on evenings and weekends are gone, at least officially. The proverbial mom-and-pop operation has passed into history. Today entry into the industry is restricted to firms with considerable investment capital.

Even though recent legislation has left the entry costs for surface operations considerably lower than start-up costs for underground mining, the former are by no means insignificant. Environmental impacts of the surface operations must be detailed precisely, accompanied by appropriate engineering studies of all reclamation activities. Approval to surface mine a tract of land is not automatic, and it may require a public hearing and several levels of governmental adjudication. In short, entry into surface operations is considerably more difficult now under the federal act and with state primacy than it was just a few years ago.[16]

Because entry into underground mining takes longer and is more difficult than entry into surface mining, the underground operations are less sensitive to changes in coal prices than the latter. In other words, an increase in the price of coal is more likely to spur additional surface output than it is to stimulate underground production. Similarly, a falling price of coal is more likely to precipitate surface mine closings. Deep-mining operations enjoy greater stability in output than do surface mines.

In summary, ease of entry into the production process is more likely to exist in surface than in underground mining. A decline in the share of coal surface mined is likely to result in reduced competition in the industry and in reduced responsiveness of producers to sudden surges in demand.

There are several other factors that frequently have an impact on the competitiveness of the coal industry. The existence of economies of scale means that as coal-mining operations are expanded, costs per ton decline. The prospects of reaping the benefits from economies of scale are much greater in the western and midwestern United States than in Appalachia. In the former regions, economies of scale occur where very large beds of coal with narrow overburdens are mined on a sustained basis. In Appalachia the topography of the land and the width of the seam are not amenable to large-scale surface mining, and hence the potential economies of scale are limited. It is the nature of the coal deposits that determines, to a very large degree, this potential for scale economies.

In underground mining the potential for declining production costs is also influenced by the size of the coal bed. Prior to the implementation of environmental and safety standards, small operators in some areas could mine coal just as efficiently as and occasionally even more efficiently than large operators. With the new restrictions designed to safeguard human life and environment, this is no longer the case. Entry costs into the industry have risen to a point at which many small operators are excluded. Consequently, average mine size will begin to increase, as will minimum efficient scale of operation.

The buyer of coal is principally interested in price, other things held constant. Large mining operations are able to take advantage of economies of scale associated with the shipping of their coal. Volume discounts and the unit train method of transport can reduce the delivered price for such operations significantly and thereby enhance their competitive advantage over smaller firms. This is further evidence that average firm size is likely to expand in the future.

Reserve Ownership

Table 18 shows by industry group and tonnage the percentage share of coal reserves held by U.S. companies domestically and in Canada. The data show that the oil companies control two-fifths of all reserves in excess of 1 billion tons and the railroads, the second largest industry group, one-fifth. These two groups, along with the coal companies, comprise the largest producers of coal. In addition to the producers are the users of coal, the electric utilities and steel companies, who likewise have an interest in holding large amounts of reserves for their future needs. Utility holdings, although large, are not dominant, and most reserves are controlled by suppliers.

Suppliers may, of course, be tempted to behave anticompetitively, for example, by withholding fuel from the market in order to create a shortage and drive up its price. Congress and some policy analysts have been concerned in the past with the impact of horizontal integration in energy industries, and in the late 1970s legislation was introduced to counteract this trend. So far no evidence has been generated to support the assertion that horizontal integration in fact has resulted in higher energy costs.

The data in table 18 also show that the major portion of the reserve tonnage, more than three-fourths, is controlled by relatively few firms. The remainder of the reserves, 22.8%, are controlled by 393 firms. This would suggest that the reserve holdings of less than 1 billion tons are indeed dispersed widely.

III
DEMAND FOR APPALACHIAN COAL

Energy resource allocation, the subject of the previous chapter, was discussed in basic theoretical terms. This chapter and the next deal specifically with Appalachian coal. They attempt to give the reader an in-depth knowledge of how the market for Appalachian coal has been affected by political and sociological events as well as by market forces. After a brief preview of the overall market for Appalachian coal, we will look extensively at the two forces which act together to determine Appalachian coal sales (or purchases) and prices. These two forces are, of course, demand and supply. Demand will be examined in the latter pages of this chapter; supply, in chapter 4.

Some allege that the market for coal spans this nation and includes other nations. Likewise, the claim has been made that many energy resource firms own subsidiaries in other fuel resources that "compete with each other throughout the United States."[1] More realistic, however, is the observation that the markets for a type and quality of coal seem to be circumscribed by the costs of transporting the product. For example, sales of coal from Eastern Kentucky are limited to an area with a radius of approximately 500 miles surrounding the producing region. Only coal shipped by barge, a less costly means of transport than rail, is sold at points more distant. Professor George Stigler, 1982 Nobel laureate in economics, defines a market as "the area within which the price of a commodity tends to uniformity, allowance being made for transportation costs."[2]

Measuring first the similarity in supply patterns among consuming states and second whether these states are substantial consumers (demanders) of coal from common supply districts, Professor Shrieves has identified several multiple-state and one single-state market (Alabama) for Appalachian coal.[3] The multiple-state markets are:

Northeastern	*Central*	*South Atlantic*
New York	Indiana	North Carolina
Pennsylvania	Wisconsin	South Carolina
Delaware and Maryland	Illinois	Virginia
West Virginia	Georgia and Florida	Missouri
New Jersey	Kentucky	Tennessee
Michigan	Iowa	

A single-state market is one in which the state consumes most of the coal it produces. With the exception of Iowa and Missouri, all of the states included in the markets delineated by Shrieves are located east of the Mississippi River. The predominant supplier by far to these states is the Appalachian coal region, which in 1984 produced and sold 439 million tons, 54% of national output.

Not all of the coal consumed east of the Mississippi River and in Iowa and Missouri originates in Appalachia. Most of the output from the Eastern Interior Coal Basin—Western Kentucky, Indiana, and Illinois—and some from the Rocky Mountain states also is consumed east of the river. But the Appalachian basin is by far the largest supplier of coal in the East, and it is the only exporter to overseas markets.

More than one-half of the bituminous coal mined in the United States originates in Appalachia. Table 19 shows the percentage share of the three bituminous-coal-producing regions from 1960 to 1984. The data illustrate the dramatic rise in the production of coal from the western United States. In the past ten years this region's output as a percentage of U.S. output increased nearly fivefold at the expense of the other two coal regions. Because western coal lies in wide seams located close to the surface, it is considerably less expensive to mine. The coal is also blessed with low concentrations of sulfur. Even though this coal has relatively low heating value and is often far from its markets, the delivered cost per Btu remains attractive in many regional markets. Gradually over the past ten years this coal has captured an increasing share of the coal markets east of the Mississippi River. In 1980 it supplied one-fourth of U.S. output, more than a tenfold increase in its U.S. market share in twenty years.

As table 13 illustrates, the northern states of Wyoming and Montana are most responsible for the dramatic increase in the region's coal output. Between 1970 and 1984, Wyoming coal production rose from 7 million to 128 million tons annually, and Montana's grew from 3.5 million to 32.8 million tons. Most of the coal moved east to the North Central and Midwest regions of the nation, markets previously considered the domain of Appalachian and interior basin producers.

It is unlikely that western coal will become a widely used fuel in the eastern United States in the near future, however. High transport costs would make its delivered price noncompetitive with that of Appalachian and interior basin coal. Thus, the market for eastern coal is expected to remain intact as long as coal can be produced at a price which is less than the delivered price per Btu of western coal.

Such a dramatic change in price structure is not expected to occur soon, and it is reasonable to expect that the demand for Appalachian coal will continue because of the needs of electric power companies, foreign buyers, coke plants, industrial sectors (including transportation), and residential and commercial consumers.

Appalachian coal is in strong demand because of its proximity to markets and because of its properties. The heat content of the coal ranges from 10,200 to 15,600 Btu per pound. Sulfur content spans a range of from 0.4%-14.0%, although most of the coal mined and sold in the 1980s has contained concentrations of sulfur in the 2% range or below. The ash content of Appalachian coal is between 2.0% and 28.0%. Of the states in the Appalachian basin, only Ohio produces coal with sulfur levels in excess of 3%. In general, northern Appalachian coal, from Ohio, Pennsylvania, and northern West Virginia, is of poorer quality than central Appalachian coal, from Eastern Kentucky and southern West Virginia.

Although the heat content of coal measured in Btu has traditionally been the most important quality variable in determining its demand, sulfur and ash content have become equally important considerations in recent years. The continuing nationwide concern for maintaining clean air seems to ensure that the quality of coal will remain important in the future as well. There are no indications that national or local air quality standards will deteriorate soon. On the contrary, concern with the effects of "acid rain" on the environment is mounting, and existing sulfur and ash emission standards may actually be tightened by congressional legislation now under review. If this occurs, the demand for clean, high-grade central Appalachian coal may surge dramatically and with it, its price.

The best measure of the demand for Appalachian coal is the distribution of coal shipments by user group. These shipments reflect the desires of different users to buy coal for combustion, for inventory adjustment, or for both purposes. Inventory adjustment is important only in the short run, in which real or expected disturbances may influence buyers to alter their coal stocks. In the long run the size of stocks is pretty much a constant function of the amount of coal burned in a given time period.

As expected, the largest group of consumers of Appalachian coal continues to be the electric utilities (see table 20). Between 1970 and 1984 their consumption of Appalachian coal increased by 40%, a robust average annual rate of 2.9%. The sharpest increase occurred after the oil-boycott years of 1974 and 1975 when electric utilities pressed into service as many coal-fired furnaces as possible. After that sudden surge in the electric utilities' demand for Appalachian coal, the annual increase returned to more conservative levels.

The most remarkable change in the consumption of Appalachian coal is that in the total amount of coal consumed. In 1982 the same volume of Appalachian coal, 410 million short tons, was consumed as in 1970. Because total U.S. coal production rose, however, Appalachia's share has declined steadily, from 71% of national output in 1970 to 54% in 1983 and 1984 (see table 19).

While the total average volume of Appalachian coal sold remained fairly constant between 1970 and 1982, the electric utilities have significantly increased their share of the amount purchased. In 1982, 58% of all Appalachian coal mined was shipped to electric utilities, considerably more than the 47% of 1970 output. In 1984 the share rose to 62%. For this region the electric utility market in the United States and abroad appears to have been the only expanding market in the past fourteen years. The other two sets of domestic consumers in table 20, coking plants and the "other" category, which includes retail, industrial, commercial, and transportation users, have dramatically decreased their use of coal. In part this decline is a reflection of worldwide excess capacity to produce steel and of vigorous foreign competition in the sale of unmanufactured steel. In part, also, it is a reflection of advances in steel-making technology that have steadily reduced the amount of coke required for the process. Between 1970 and 1982, the shipments of coal to coke plants declined by nearly two-thirds. The most noteworthy decline occurred in 1982, when the deep recession sharply reduced the demand for consumer products, the demand for steel used to make such products and, of course, the demand for coking coal. In that year alone shipments to coke ovens declined by nearly one-third.

The reduction in demand by retail consumers and industrial, commercial, and transportation users continued its decline through the

1970s and into the 1980s. Consumption in 1982 by those users was only one-half of the 1970 level. The most precipitous decline came between 1970 and 1975 as conversion efforts from coal-fired to oil-fired equipment begun in the 1960s came to fruition. The oil crisis of 1973 put a temporary halt to some of the conversion plans, but if the real price of oil continues to fall in the 1980s, the trend is likely to continue.

THE ELECTRIC POWER DEMAND FOR COAL

In 1984, 664 million short tons of coal were burned in the United States to generate 56% of the electricity consumed by the nation.[4] That represented nearly 84% of all the coal consumed in the nation that year, a ratio that has been rising steadily since at least 1970.

Of the 664 million tons, 271 million, or 41%, were supplied by Appalachia, an important and also steadily increasing share. Some of the coal shipped to the electric utilities actually may have been stockpiled because the severity of the recession reduced the industrial demand for electricity rather significantly. Such consumer stock adjustments are unlikely to have exceeded 4 million tons for the year and probably were considerably smaller.

For most of the century the growth in the demand for electric power has been greater than the growth in real gross national product. At the same time, since the 1920s the ratio of total energy to real gross national product (the energy/GNP ratio) has been declining, with only two years providing exceptions. The principal explanations for these trends are the increased thermal efficiency of energy use, the effects of widespread electrification, and the general rise of productivity in the economy. The energy/GNP ratio declined following World War II through 1966, then rose for a few years, but the fall resumed following the oil-crisis years of 1973-1974. The reasons for the abrupt 1967 trend reversal are the slowdown in the increase in productivity and the substitution of electric power for direct fuel use in residential, commercial, and industrial heating, an inferior but more convenient use of energy resources. These trends in electric power consumption were exacerbated by a persistent decline in the relative price of electricity. If these prices do not rise as rapidly as the general price level, or if they decline as they did for all of the 1960s, consumers will have an incentive to expand their use of electric power. Prices were steady or in decline up to 1974, the year in which the electric utility companies began to pass on rising fuel costs in the form of higher electricity prices.

Table 21 shows the price behavior of electricity by end user for the past twenty-four years. The average price of electric power used in commerce and by private residences showed a remarkable degree of stability over the 1960-1972 period. Residential electricity prices declined by 13%, commercial prices by only 6%. Industrial electricity prices showed greater volatility and fell by 25% during the period.

For the next ten years, however, the real price of electric power rose sharply. By 1984 it had doubled for industrial consumers and had increased by 42% for private residential customers and by 45% for commercial establishments. The principal source of these rate escalations was the 1973 oil boycott and the subsequent rise in energy resource prices. Another source was the haste with which many electric power companies entered into long-term contracts with fuel suppliers at uncommonly high prices. As soon as it became permissible to pass to the consumer the higher costs of energy fuels, public utilities lost most of their incentive to seek least-cost suppliers. Instead, they negotiated long-term contracts for oil, coal, and gas, placing foremost attention on the reliability of future supplies and only secondary emphasis on price. Table 5 also shows that the delivered cost of coal measured in cents per million had risen far less sharply than the equivalent cost for petroleum and natural gas.

All of these factors, but primarily the question of supply reliability, naturally shifted attention away from oil and toward coal as a future energy source. The unpredictability of international events and tensions and their impact on oil exports gradually convinced domestic buyers that, for the 1970s at least, U.S. coal was an unimpeachably secure energy source. To the extent of technological feasibility, increasingly more kilowatt hours of electricity began to be generated from coal. Table 22 shows the steady growth in coal-generated power in comparison with that of petroleum- and gas-generated power. In 1984, electricity produced from coal had risen to 1,342 billion kilowatt hours, a 74% increase over 1972. During the same period petroleum use declined by more than one-half, and natural gas use declined by 21%. In 1984 more than four times as much electricity was produced from coal than from either natural gas or nuclear resources. Particularly remarkable is the decline in the use of petroleum in the electric utility industry. This decline started in 1973 and, with the exception of three years, continues to the present. In 1984, petroleum-based electric power accounted for only 120 billion kilowatt hours, almost one-third of the amount produced by nuclear facilities. Petroleum has become the least important of the five major sources of U.S. power output. This dramatic transformation of the role of petroleum fuels reflects the underlying principle of substitution of less expensive for more expensive energy resources. The recessions of the 1980s have enhanced this transformation, with oil-fired boilers being idled before coal-fired ones when reduced production has been necessary.

Generally it seems that the energy sources for the production of electric power are well diversified in the United States. More than 55% of the electricity produced in 1984 originated from coal, the most abundant source of fossil fuel energy in the nation. Hydropower, natural gas, and nuclear resources were used about equally in the production of power, while petroleum placed a distant fifth. The nation is now much less dependent on oil, domestic or imported, than it once was. In 1982, the year of deepest recession and the first since the depression during which electricity output actually declined, only 6.5% of the total kilowatt hours of electricity produced in the United States had their origin in petroleum.

The largest coal-producing district in Appalachia is district 8, which includes Eastern Kentucky, North Carolina, Tennessee, and portions of West Virginia and Virginia. In 1982, close to one-half of the coal mined in the Appalachian basin originated in this district. Table 23 exhibits by consumer use the destination of district 8 coal and of the other Appalachian coal for 1984. The uses of Appalachian coal from both origins appear to be quite similar. Sixty-three percent of the Appalachian coal mined is shipped to electric utilities. The next largest demand source is foreign, including Canada and Mexico. Coke ovens and the commercial, retail, and industrial categories absorb the remainder.

Because the demand for coal is derived from the demand for electric power, it is important to examine the nature of U.S. and regional demands for electricity and its fuel sources. The information provided in table 24 shows that, with the exception of the New England and Middle Atlantic census regions, which include the eastern states from Maine to Pennsylvania, all other regions in the United States expanded their production of electric power between the years 1972 and 1982. This expansion commenced in 1972 and proceeded uninterrupted through 1982. Only the Pacific region shows a decline in electric power production from 1977 to 1981. In the Middle Atlantic region, production of electricity remained remarkably stable: 181.8 billion kilowatt hours in 1972, and 179.2 billion hours in 1982. For the New England region, the decline in power production for the ten years was only 14%, not a dramatic change.

In contrast, the sharpest growth in power output occurred in the eight Mountain region states, which range from Montana to Arizona.

In 1982, two and one-fourth times as much electricity was generated there than ten years earlier.

In nearly every U.S. census region listed in table 24, coal has increased its share of the electric utility market in comparison with the other two fossil fuels. In percentage terms, the smallest increase occurred in the East North Central and Pacific regions; the largest occurred in New England. Also very large was the increase in the West South Central region, which includes Arkansas, Louisiana, Oklahoma, and Texas. With Texas's large deposits of lignite and subbituminous coal, that state's electric utilities have become the largest consuming sector for U.S. coal. The next largest group of state utilities includes those located in Ohio, Indiana, Illinois, and West Virginia, all coal-producing states.

The gradual switch to coal took place against losses in utility market shares by oil and gas. As table 24 illustrates, since 1977 oil has lost in market share in every region of the nation. Between 1972 and 1977 oil increased its market share despite the rapidly rising price of oil. The switch to alternate fuels took time, because the substitution effect of higher oil prices is primarily a long-term phenomenon. Between 1977 and 1981 oil was displaced at a rapid pace.

Natural gas also lost some of its appeal as a boiler fuel between 1972 and 1982. Shortages at governing prices became widespread, and natural gas reserves were said to be dwindling rapidly. With the exception in the Middle Atlantic and Pacific regions, gas experienced a sharp decline in market share throughout the United States.

The largest portion of the electric utility demand for Appalachian coal originates in the Middle Atlantic, South Atlantic, East North Central, and East South Central regions of the nation. These four regions consumed a total of 364,281 tons of coal in electricity production in 1982;[5] this figure represents more than 60% of the total being burned by U.S. utilities. Of the four, the East North Central region (containing Illinois, Indiana, Michigan, Ohio, and Wisconsin) consumed 186,552 million tons. Of course, not all of this coal was mined in Appalachia. It is certain that Indiana and Illinois used large quantities of their indigenous coal. Because of the high sulfur content of these states' coal, however, it is frequently blended with Appalachian coal. Ohio relies to a very large extent on its own coal sources, while Michigan and Wisconsin buy coal from both the eastern interior basin and from Appalachia.

In (descending) order of coal consumption, the East North Central region is followed by the South Atlantic region, West North Central region, and East South Central region. The largest coal-consuming state in the four regions was Ohio, the smallest, Delaware.

Along the Eastern Seaboard, coal and oil have been the principal fuel sources for the generation of electric power for some time. Some states, such as New Jersey, Rhode Island, Florida, and New York, use sizable amounts of natural gas for the production of electricity, but for the most part coal and oil are the most widely used fossil fuels in the East.

The major determinants of a public utility's long-term fuel choice are the delivered cost per Btu of each fuel, the reliability of delivery, and the costs involved in making the burning of each fuel acceptable for clean-air standards. Because the cost of interfuel substitution (that is, of switching from one fuel to another) is extremely high, a temporary change in relative fuel prices does not produce much substitution. It would take a major, enduring change in relative fuel prices to alter consumption patterns.

Of the three fossil fuels, natural gas is the cleanest and the least expensive to burn, but its use by electric utilities is limited. Prior to 1974 natural gas provided a reliable fuel source, but after that year supply interruptions to electric utilities became more frequent, mostly as a consequence of federal pricing policies. Until very recently the price of natural gas sold in interstate commerce has been kept below market-clearing levels by the U.S. Federal Power Commission (FPC). Instead of promoting balanced development of demand and supply, prices were set at levels for most gas that stimulated demand and discouraged supply.[6] One result was the increased occurrence of interruptions of natural gas service to electric utilities and selected commercial customers as suppliers cut back production or supplied the unregulated intrastate market first. Subsequent developments forced Congress to reappraise the situation and pass legislation designed essentially to deregulate the price of natural gas by 1985. In 1983, natural gas supplies were abundant. The recession years of the 1980s, sharply higher prices of natural gas, and widespread conservation have had a predictable impact on the demand for this fuel: it has declined significantly.

The strongest competitor for coal's share in the electric utility market is residual fuel oil. This is a thick, tarlike substance that remains when crude oil is distilled into lighter products such as gasoline or diesel oil; it is generally low in sulfur. Historically, domestic oil refiners have found it profitable to refine as much of the crude oil as was technologically feasible and to leave as little residual as possible. Foreign producers, on the other hand, have traditionally produced larger quantities of residual oil.

Until 1966 oil was protected from foreign competition by a strictly enforced set of import quotas on foreign residual oil. In that year, however, the quotas were lifted on imports into the region known as Petroleum Administration for Defense District 1, which included the thirteen Atlantic Coast states and Pennsylvania, Vermont, and West Virginia. As a result, oil began to be substituted for coal along the Eastern Seaboard. Table 25 illustrates this very clearly for states such as Massachusetts, Rhode Island, Connecticut, New York, and Virginia. In fact, all states except New Hampshire significantly increased their use of oil at the expense of coal between 1966 and 1972.

The prospects of still stricter sulfur emission standards at first seemed to assure a continuing contraction of coal's share in the electric utilities market, but the oil embargo of late 1973 and the ensuing escalation in oil prices raised serious doubts in the minds of utility officials concerning the wisdom of their oil policies. Some, such as officials in New Jersey and Maryland, found it possible to reduce their reliance on oil by 1977; others did not. In some locations—New York, Pennsylvania, and South Carolina—oil consumption actually increased during this period as new, previously planned, oil-fired furnaces began to be used.

By the early 1980s the picture had changed dramatically. In nearly all states listed in table 25, coal has begun to recapture a significant portion of the electric utility market. The exceptions are Maine, Connecticut, and the District of Columbia, which now burn oil exclusively. Only New Hampshire appears to be maintaining a near balance between the use of oil and coal as energy sources.

The discussion of the alternatives to coal as a fuel for generating electric power would be incomplete without the mention of nuclear power. As of March 1985, eighty-nine nuclear reactors had the capacity to produce 72.9 billion net kilowatts of electricity per month, 16% of the U.S. total. In 1984 these facilities produced 327.6 billion kilowatt hours, 13.6% of all electricity generated in the United States.[7] This was the largest share of U.S. electricity ever produced by nuclear power. It is unlikely, however, that significantly more nuclear power generation will occur in the future. Only three construction permits were pending in 1982, and only two new units were on order. Although some of the sixty units now under construction will be completed, others have already been abandoned. Once touted as the solution to escalating needs for electric power, today nuclear power is the enfant terrible of the utility industry. Controversy regarding its possible adverse environmental effects, rising construction costs, user burden, safety and technical problems, and severely scaled-down estimates of future power needs raise serious doubts about the long-term role of nuclear power as an energy source. A decade ago the Battelle Memorial Institute estimated that nuclear power would supply as much as 50% of our electric power needs by the year 2000, and that most of this power would come from second-generation nuclear reactors known as breeder reactors.[8] Today such predictions would be, to say the least, seriously questioned.

For some states, nuclear power has become a surprisingly important source of electric energy. For example, as shown in table 26, in Vermont 77.5% of the electricity produced in January 1985 came from nuclear reactors. For South Carolina, Connecticut, and Maine, the percentages were 55.7, 54.7 and 64.3, respectively. In general, during the months of a deep recession period, electric utilities prefer to use first their high-capital-cost equipment and to use it as intensively as possible. Consequently, the nuclear components show up more strongly as generators of electricity during such periods than otherwise would be the case. Still, the percentages are remarkably high, particularly for such coal-producing states as Illinois, Alabama, and Tennessee.[9]

Except for the Middle Atlantic region, the five regions listed in table 26 show an increased share of nuclear power as an energy fuel in 1982 and 1985. In part the reason for this is the recession and its inevitable impact on the demand for electric power. In part also, the higher share reflects the existence of substantial nuclear capacity to produce electricity in the East. Even so, the prospects of continued expansion of nuclear power are poor. As reactors wear out, are dismantled, and are not replaced, the potential for expanded coal and oil use in future decades is naturally enhanced.

Since 1970 the electric utilities located east of the Mississippi River have steadily increased their consumption of Appalachian coal. In that year they purchased 47% of the basin's output; in 1984, 62% (see table 20).

Of the Appalachian areas, through 1982, Eastern Kentucky had become the largest single supplier of coal to electric utilities (see table 27). In the past ten years its shipments to power plants doubled, while those from West Virginia, Virginia, and Pennsylvania increased only modestly. Ohio's shipments actually declined, mostly because its coal contains larger quantities of sulfur than that of its neighboring states. Although in 1982 the gap narrowed somewhat, in 1981 Eastern Kentucky mines shipped half again as much coal as the nearest competing state region. In 1983, however, West Virginia shipments were 6% higher than Kentucky's. Eastern Kentucky's coal is low in sulfur and high in ability to produce heat, and traditionally it has not been hampered as much by supply interruptions as coal mined elsewhere in Appalachia. Because the Eastern Kentucky labor force is less unionized overall than labor in surrounding states, mine operators often have realized profits from labor strikes held in other states. During the last two strike years, 1978 and 1981, Eastern Kentucky increased its shipments to electric utilities while coal shipments from the other Appalachian states declined.

In addition to competing with other Appalachian coals on the basis of quality, Eastern Kentucky coal also must compete on the basis of location. This is, of course, due to the impact of distance on transportation costs and therefore on delivered price. It is not unexpected that public utilities located in New York and New Jersey consume only a small amount of Eastern Kentucky coal; Pennsylvania, contiguous to these two states, supplies the bulk of the steam coal needed there. Districts 3 and 6 (northern West Virginia) satisfy another large portion of the demand. Eastern Kentucky captured only 8.5% of the utility market in New York in 1981 and only 5.0% in New Jersey.

Table 28 lists the quantity of total Eastern Kentucky coal shipped to electric utilities in the four major consuming regions, shows the percentage of total shipments from Kentucky received by each state, and illustrates the degree to which each state depends, in terms of market share, on Eastern Kentucky coal. For example, in 1983 South Carolina utilities received 6.2 million tons of coal from Eastern Kentucky. This represented 10.0% of Eastern Kentucky shipments to utilities but was 81.9% of the coal burned by South Carolina utilities that year. It is clear that South Carolina depends on Eastern Kentucky coal. Kentucky shippers, on the other hand, face dispersed markets: no single state receives more than 16% of the coal mined for electric utilities. This dispersion represents a degree of security to mining operators, because even if one or more state markets were lost, the overall impact on the Eastern Kentucky coal industry would be small.

By far the largest proportion of Eastern Kentucky utility coal is shipped to two regions: the South Atlantic and the East North Central. North and South Carolina, Georgia, Michigan, Ohio, and Kentucky purchased 47 million tons of Kentucky steam coal in 1983, more than three-fourths of total utility coal production. Table 29 ranks in descending order the states that consume Eastern Kentucky utility coal. Kentucky and Georgia are the two states that bought the largest amounts of Eastern Kentucky coal in 1983. These two states were also important consumers in 1972.[10] It is noteworthy that Michigan has become the third most important market for Eastern Kentucky utility coal, despite its distance from the source. In an environmentally conscious world, however, this fact is perhaps not all that surprising. Clean Eastern Kentucky compliance coal (coal that meets clean-air standards when burned) can easily be blended with other, less attractive coal. In Ohio, electric utilities presently blend indigenous higher-sulfur coal with Eastern Kentucky coal in order to meet ambient air standards. But it must be remembered that both Michigan and Ohio, in contrast with the Carolinas and Kentucky, are major industrial states. The oscillations of economic activity there, generally more severe than changes, for instance, in the Carolinas, are transmitted directly to the Eastern Kentucky coal industry. As the demand for industrial and consumer goods produced in Ohio and Michigan declines, so does the demand for electric power and with it the demand for coal. As the consumer and industrial demand for electricity from the Detroit Edison Company in Michigan falls, so does the amount of coal needed and the amount of coal mined in Pike County, Kentucky. The linkage is direct.

Since the early 1970s the electric utilities have become increasingly more important buyers of Appalachian coal, in particular of district 8 coal. In 1970 they consumed 38% of the district's production; in 1983, 59%. Table 30 shows the growth in the utility and in the export demand for district 8 coal. In 1970 these two demand sectors purchased 60% of the coal mined in the district; in 1983, 79%. It is clear that these two sources of demand influence strongly the fortunes of the entire Appalachian coal industry and of the people who live there.

The increase in the demand for Appalachian coal, and in particular for central Appalachian coal, which includes district 8, is ascribable to a very large degree to the passage of the U.S. Clean Air Act of 1970. This legislation set national ambient air quality standards and charged the states with responsibility for enforcement. The states in turn established maximum sulfur dioxide (SO_2) emissions levels for existing utility power plants. New coal-burning power plants, those on which construction was begun after August 17, 1971, were required to meet an SO_2 standard of 1.2 pounds per million Btu. This legislation was amended in 1977 with new standards for plants constructed after September 1978, but as of 1981 no such plants had been put into operation.[11]

Initially, and in anticipation of these air quality standards, some utilities began to substitute oil for coal in power generation. The 1973 oil embargo and the subsequent upward spiral in oil prices put a sudden halt to that; electric utilities began to reassess their plans, and some returned to coal as a primary fuel.

The electric utilities basically had two alternatives for meeting clean air standards. They could purchase flue gas desulfurization (FGD) equipment, which removes the sulfur dioxide from the stacks after combustion through a process called scrubbing, or they could use compliance coal, which after combustion emits less than 1.2 pounds of sulfur per million Btu produced. Because the capital costs of scrubbing equipment are very high, the technology not reliable, and maintenance and sulfur disposal costs considerable, and because expenditures on new capital equipment can be recaptured only with public service commissions' approval, most utilities chose the second alternative. They considered it cost effective to purchase compliance coal, even though they realized that the price of this coal would eventually increase disproportionately to the prices of other types of coal. This prospect did not concern them greatly because under the fuel adjustment clause of the 1970 legislation they were able to pass forward to the consumer the higher price of coal. Because Eastern Kentucky contains large deposits of low-sulfur compliance coal, the demand for it began to rise steadily. By 1981 this region shipped 50% more coal to utilities than did the nearest competitor. Not only did existing plants burn more compliance coal in the 1970s, but also new plants preferred this alternative to the costly addition of FGD equipment.

Of the sixteen states whose electric utilities rely heavily on Appalachian coal, eleven buy coal exclusively from the region and another three purchase at least 90% of their coal from Appalachia. Table 31 shows the 1982 volume of Appalachian coal shipped to its utility customers and their dependence on this coal. Illinois and Indiana use mostly indigenous coal, as do many Kentucky-based utilities, particularly those in Western Kentucky. In fact, in 1982 the origin of steam coal consumed by Kentucky electric utilities producing 25 MW or more electricity was as follows:

Area of Origin	Percentage
Eastern Kentucky	34.0
Western Kentucky	48.4
Appalachian states except Kentucky	10.7
Eastern interior basin except Kentucky	6.9
Total	100.0

The distribution of sources shows that 55% (48.4 plus 6.9) of that coal originated in the high-sulfur eastern interior basin. Some of this coal was blended with Appalachian coal to meet emissions standards, and other coal was burned as is, subject to sulfur removal with FGD equipment.

Michigan utilities, which in recent years have become important consumers of Eastern Kentucky coal, also buy a large amount of their coal from the western United States, principally Wyoming. In the past ten years Michigan has dramatically reduced its purchases from the other Appalachian states and increased significantly its consumption of Eastern Kentucky coal. This suggests that Michigan purchases are a clear example of the competitive inroads western coal has made into traditional Appalachian markets. For Indiana, Illinois, Wisconsin, and Mississippi, the story is pretty much the same. In Mississippi, on the other hand, it is interior basin coal that is being displaced by coal from western suppliers, with Appalachia's share remaining relatively unchanged.

Although not a major coal-consuming state, Florida houses utilities that have experimented with the use of many different types of coal from all over the world. But the state's utilities also seem to be increasing steadily their consumption of Kentucky coal. In 1981, 54% of the coal consumed in the state was mined in Kentucky; this figure was 14% in 1973. In 1978 and 1979 Florida purchased coal not only domestically but also from South Africa, Poland, and Australia. In 1981 only South Africa still supplied coal to Florida, and then only about ten shiploads, 7% of total consumption. It appears that foreign coal is unlikely to be a serious competitor on U.S. shores in the future. Only Florida, Alabama, and Mississippi have experimented with burning such coal, and they have now mostly abandoned their efforts. Even though these foreign coals produce sufficient Btu and have low enough sulfur content to meet utility specifications and U.S. environmental standards,[12] in the minds of most utility buyers the long-term reliability of supply is uncertain. By contrast, the long-term reliability of U.S. exports is never in question. Foreign buyers also perceive the United States as the most reliable, though not the least expensive, supplier of quality coal. The act of buying it strongly reveals their preference. The price differential of as much as fifteen dollars per ton delivered in Europe reflects European buyers' willingness to pay extra for a secure supply and for diversification of supply sources.

EXPORT DEMAND FOR COAL

For many years the United States exported only the highest-quality and most valuable coal mined, mostly metallurgical-grade coal destined for use in the making of coke. As table 32 illustrates, this has changed over the past ten years. In that time exports doubled, with the greatest surge occurring during 1980-1982. Although exports of metallurgical coal increased by 50% in ten years, steam coal shipments rose by 300%. As recently as 1978 only 10 million tons of steam coal were exported to foreign utilities. By 1982 that number had risen to 41 million tons. In the span of these four years, steam exports increased by a factor of four; metallurgical exports increased to twice their 1978 level.

Like all other commodities that are traded internationally, coal is affected by the value of the dollar abroad. The relative values of international currencies take on particular importance in determining trade flows when commodity markets are weak, when ample supplies are offered for sale and buyers are scarce. This was the case in the international coal market in 1983 and 1984.

Astounding, at first, is the fact that despite an increasingly strong dollar, U.S. coal exports continued to surge upward from their 1980 level. Despite a rise in the price foreign buyers were paying for the dollar and therefore for coal, increasing quantities of U.S. coal were being demanded. The explanation for this unusual situation is found in the dramatic reduction of coal production in exports from Poland and in the impact of labor unrest on Australian export.[13] These serious supply interruptions induced European and Japanese buyers to seek coal elsewhere. With South Africa already producing at capacity, U.S. coal operations provided the only alternative. As table 21 shows, Appalachian coal exports rose by over 25 million tons in 1980 alone, an increase of 55% over the preceding year, and the export share of total Appalachian shipments increased from 11% to 17%. The export boom continued into 1982 but began to weaken late in the year. Increasingly, the overvalued dollar, the worldwide recession, and the removal of supply obstacles in Poland and Australia had their predictable effect on U.S. exports. Even so, Appalachian exports have remained remarkably strong, probably as a result of the high reliability of U.S. supply flows. Eventually, however, if the overvaluation of the dollar persists and other factors remain unchanged, foreigners will buy increasingly less coal from the United States. If future monetary and fiscal policies are unsuccessful in effecting a reduction in the value of the dollar, the market will seek its own solution: reduced exports.

As table 33 illustrates, exports in 1982 declined from their level of the previous year. This decrease was most pronounced in end-of-year shipments to Europe. The recession there, warmer-than-usual temperatures, bulging stockpiles, and vigorous competition from Polish coal combined to depress final-quarter shipments in 1982. Demand in the Netherlands, Belgium-Luxembourg, West Germany, Spain, Denmark, Yugoslavia, and the United Kingdom declined precipitously from levels in the last quarter of 1981. The 45% decrease was only marginally offset by increased demand from France, Greece, Turkey, Norway, and Sweden. The decline in exports accelerated significantly in 1983 and flattened out in 1984.

The largest single buyer of U.S. coal is Canada, followed by Japan and Italy. These three markets together absorbed 54% of 1984 U.S. exports. Table 34 shows the principal buying countries of U.S. coal listed in declining order of tonnage received.

The type of coal going to each market varies greatly (see table 35). Most of the U.S. coal going to Japan is of metallurgical grade for use in the steel industry. In 1983 only 1.7 million tons were destined for Japanese electric utilities. Exports to Canada, on the other hand, consist mostly of steam coal, although almost 7 million tons were used by the steel industry. Japan's 1983 purchases made up about 32% of U.S. metallurgical coal exports, and Canada absorbed 37% of total steam coal exports—both sizable portions of the export market. In both cases, American coal exports are closely related to economic conditions in these two countries. Because these two countries export a significant portion of their output to the United States, they in turn depend on the level of economic activity here. During depressed times in the United States, Japanese and Canadian demand for U.S. coal weakens. Thus, the vigor of the U.S. economy influences directly the domestic demand for coal, as well as indirectly the foreign demand for coal.

Sizable shipments of U.S.-produced steam coal find their way to Europe. In 1983, 13.4 million tons crossed the Atlantic, a reflection of the increased interest of electric utilities in American coal because of the delays and uncertainties that characterized Australian, Polish, and South African coal shipments. In part, also, Europe has begun to lessen gradually its dependence on oil as an energy source. Despite high mine-to-port rail costs and sizable ocean transport charges, U.S. coal competed successfully in 1981 and 1982 with other coals in European markets, particularly those mined in Belgium, Germany, and England. On the continent mining conditions are not very favorable and production costs are high. Even with government subsidies European coal remains uncompetitive with imported coal. To a lesser extent this is also true of England's coal.[14]

Given the location of U.S. coal deposits and mining activities, it is not surprising that nearly two-thirds of 1983 coal exports flowed through East Coast ports. Table 36 details ports of loading for coal exports. These ports can be grouped into four distinct regions geographically. The Northern Great Lakes region generally handles only coal for export to Canada, whereas the other three regions service overseas shipments.

The East Coast ports, in particular Norfolk, Virginia; Baltimore; and Philadelphia, in declining order of volume, handled 99% of the coal export traffic from Appalachia. In 1983, 40.7 million tons of coal were loaded onto ships at the various docks in Norfolk. That is exactly the entire amount of coal exported from the United States in 1978. Exports moving north to Canada flow almost entirely through Cleveland. That lake port alone handles 99% of the exports to Canada; similarly, Los Angeles, the dominant West Coast port, handles 94% of exports to points west of the United States. On the Gulf Coast there are two major port facilities: New Orleans, being the largest and Mobile, Alabama, the second largest.

In 1983 the United States exported 76.9 million tons of coal. Appalachian coal producers provided 72.3 million tons, or 94% of this coal. More than one-half of the Appalachian coal exported originated in district 8, making it the largest coal-exporting district in the nation by far.[15] Despite the worldwide recession, which affected the demand not only for steam coal but for metallurgical coal as well, district 8 overseas exports have remained high each of the past four years. In 1982 these exports, at 40.7 million tons, were twice as large as in 1979.

Historically, it has been Appalachia that has provided virtually all of the coal exported from the United States. Within the Appalachian Coal Basin, central Appalachia, which includes district 8 and Eastern Kentucky coal, exported 11.8 million tons in 1981 and 9.7 million tons in 1982. Table 37 shows the contribution of the three coal-producing regions to the export market. Recent Japanese interest in western steam coal explains that region's growth in the 1980s, but in Europe and Canada only Appalachian coal is in demand. Very much like Western European purchasers, Japanese buyers of Australian coal have become concerned over the unreliability of deliveries from that country from time to time and have turned to U.S. suppliers.

At the heart of foreign demand for coal is each nation's generally high use of energy and an economy that is affected strongly by the level of economic activity and by relative changes in fuel prices. Very much like in the United States, the general rise in fuel prices in these countries has led to widespread measures to conserve energy use; it has also led to interfuel substitution.

In the principal nations of the Organization for Economic Cooperation and Development (OECD), the largest consumers of energy are in the industrial sectors, which used about 41% of all energy produced during the period 1960-1973. But following the sharp oil price increases, which began in Europe as early as 1969 and continued into the 1970s, energy consumption began to decline. This decline is reflected in the steady fall of the energy-consumption-to-output ratio for most OECD countries and was undoubtedly influenced by the OPEC oil price increases of 1973-1974 and 1979. The indices shown in table 38 for the major OECD countries illustrate the negative correlation between the energy-to-output ratios and the increasing real price of energy. The decline in the ratios is most pronounced in the United States, France, the United Kingdom, and Japan. It is said that because much was being used wastefully in the two English-speaking nations and France, conservation measures were most easily implemented there. In Japan, the institutional climate made the implementation of conservation procedures a relatively easy task. The most pronounced drop in energy consumption per unit of output took place in 1974, immediately after the fourfold increase in world oil prices. After that the average decline followed a relatively constant pace until 1980, when the second round of OPEC oil price increases had its predictable effect.

Figure 3 depicts the time trend of the energy-to-output ratio and of the real price of energy to industry. It illustrates that following the price increases of 1969 energy consumption per unit of output began a steady, uninterrupted decline, which continues into the present. The steep energy price increases of 1973 do not seem to have greatly affected the rate of decline of the ratios in the seven largest OECD nations, in which the process of energy conservation appears to have already been institutionalized.

During the 1960s and into the early 1970s, energy consumption in the industrial sectors of the major OECD nations grew at about 5% annually. After 1973, however, energy use began a steady decline, although the fall was most pronounced among the major member nations. The declines were steepest in Germany and the United Kingdom and more modest in France, Italy, and Canada. In Japan and the United States, industrial energy consumption did not decline absolutely, but the rate of growth fell to less than 1%, a very low rate when compared with the growth of the 1960s and the early 1970s. (In Japan a very sharp drop occurred in 1980.)

From the 1960s to 1979, the demand for electric energy grew at rates greater than the growth of total demand for energy. Even after 1979, while total primary energy consumption in the OECD coun-

Figure 3. The Energy/Output and Real Energy Prices of Industry
in the Seven Largest OECD Economies

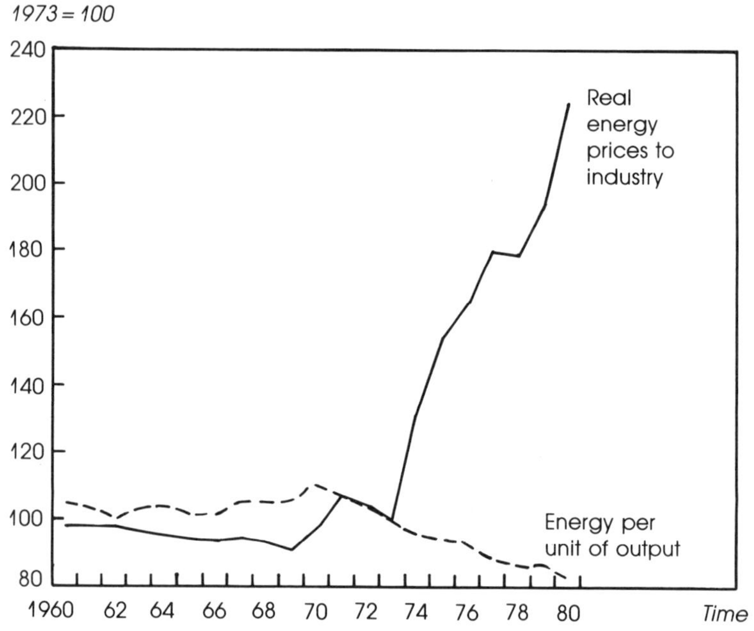

Source: International Energy Agency, OECD, *World Energy Outlook*,
Paris: OECD/IEA, 1982.

tries declined by 2.8% annually, electricity consumption rose by
1.3%. Consequently, electricity's share in each country's gross product increased, on average, from 14% in 1973 to 16% in 1980.[16]

Electricity has historically been used for special purposes. In industry it propels electric motors, assists in the plating of sheet metals,
and heats space. In the first two uses, there exist no technical
substitutes for electricity. For example, industrial robots are usually
powered by electric motors, and alternative fuel inputs seem imprac-
tical. In space heating, interfuel substitution of fossil fuels for electricity is possible but is often limited by special needs and ap-
plications.

Of considerable interest to U.S. coal exporters is the question of
how sensitive foreign buyers are to changes in the delivered price of
coal. In a study using pre-1974 time series data for ten industrialized
countries (except the United States), Professor Pindyck estimated that
the own-price elasticities for solid fuels, including anthracite and coke,
ranged from −1.29 to −2.34 in the long run. This means that a 1% in-

crease in the price of coal would result in a 1.29% to 2.34% decrease
in the quantity of coal demanded, a fairly undramatic price re-
action.[17] Whether these low-elasticity coefficients are valid ten years
later in the post-energy crisis period is uncertain, but it is clear that
price is not the most important determinant of the demand for coal
abroad. Rather, foreign electric utilities are more concerned with
maintaining a diversified network of suppliers, both among supplying nations and among supplying firms within each. They are keen-
ly interested in a secure source of supply, timely delivery, and consistent quality.[18]

In the residential and commercial sectors of each foreign
economy, the demand for electricity is derived from the use of lighting
equipment and the operation of appliances. In recent years this demand has grown more rapidly than has GNP, and this pattern is likely
to continue. Other uses of electricity are possibly more vulnerable.
Interfuel price differentials can have a significant impact on the type
of space-heating, water-heating, and cooking equipment used in the
future, and for such purposes the use of electricity may decline.

Between 1975 and 1980 residential and commercial demand for
energy increased at an annual rate of only 1% in the OECD countries. The demand for electricity, on the other hand, rose by 4.1%
annually.[19] This increase is a clear reflection of the strong link in industrialized countries between consumer spending and the demand
for electric power.

Historically, coal has served as the primary fuel source for the
generation of electric power. Between 1973 and 1982 it increased its
share in North American electricity production by about 11% but
declined in European production by 3%. In Japan, coal's share rose
by 39%, but it is still less than 20% of all primary fuels used to produce electricity (see table 39).

It is not unexpected that countries endowed with coal deposits
use more coal in industry and in the production of electricity than
countries that have no coal. The largest users of coal are the electric
utilities in the United Kingdom, the United States, and West Germany. In the first two, according to OECD statistics, coal's share has
increased significantly between 1973 and 1982. In West Germany it
has not. Nuclear reactors, natural gas, and oil have continued to re-
tain or expand their share of this market. Noteworthy also is the
threefold increase in Italy's use of coal by electric utilities. Because
the share of coal in primary fuel use there and in Japan is still relatively small, these two countries would seem to be prime candidates for
further market penetration by coal exports.

The iron and steel industry is the principal industrial consumer of coal, particularly of metallurgical-grade coal. European and Japanese demand for U.S. metallurgical coal is strongly influenced by the demand for steel. During the early 1980s the worldwide slowdown in economic activity and in the demand for steel and coke weakened the demand for metallurgical coal. Moreover, technological changes in the making of steel have gradually reduced the amount of coal needed. The continuous-casting process reduces energy requirements because molten steel is cast as it is produced. Hence, less coke is needed. Also, increasing numbers of countries are producing steel from scrap iron in electric furnaces, greatly reducing the demand for crude iron and for coke. The only compensatory influence is the trend among the countries now producing metallurgical coal—West Germany, the United Kingdom, and to a limited extent France, Belgium, and Spain—to mine less of it. Only West Germany now exports metallurgical coal, but the amount it produces is not enough to meet European demand. Any future increases in European demand for metallurgical coal, if they in fact materialize, would have to be met by increases in imports, presumably from the United States.

Another positive factor for coal is that a significant number of oil- and gas-fired furnaces will gradually be phased out; this retirement can be expected to offset partially the falling demand for metallurgical coal. Finally, an important factor for the future of coal is that in several Western European countries—Denmark, West Germany, and the United Kingdom—conversion to coal use in the cement industry is nearly complete. Because of the size of the cement industry, it represents an important future market for coal. Consequently, the demand for high-quality U.S. coal is likely to remain stable.

Other industrial sectors, both foreign and domestic, are not as easily able to convert to the use of coal as a primary fuel. Lack of storage space for the bulky fuel, high interest rates and prospects of economic uncertainties, permit difficulties, and unclear perceptions about the relative long-term attractiveness of coal and oil use are among the factors that militate against substantive departures from past practices. But the potential for the increased use of coal in industry is perhaps greater than realized, depending on the resolution of a number of constraints such as the convenience of conversion, the success of the implementation of atmospheric fluidized bed combustion (AFBC) in several regions, and other factors to be discussed later.

In industry, coal is generally used on a more modest scale than in electric power generation. Countries located in the Pacific region, in particular Japan, have a higher propensity to use coal in industry than other nations. In Germany, coal's one-fourth share in producing the energy used in industry has remained unchanged in recent years. Canada, on the other hand, has increased its use of coal in industry, while France and the United Kingdom have reduced it. Overall, no clear trend of coal use in industry is discernible from the past nine years.

With the exception of users in Denmark, and in the future perhaps Sweden, the residential and commercial sectors of the industrialized nations are unlikely to demand coal for heating use. Coal-fired district heating (DH) is technologically feasible and will be used more widely in Denmark in the near future, but elsewhere it is not being considered as a new source of space heating. In most industrialized societies coal has largely disappeared as a fuel for residential and commercial use. Only in South Korea is the use of anthracite briquettes for residential heating widespread.

THE METALLURGICAL DEMAND FOR COAL

When coal is heated in the absence of air, it is distilled and becomes carbonized. The resulting solid product is coke, which is used chiefly as a metallurgical fuel. When this fuel is added to iron ore and limestone in a preheated, pressurized air chamber, combustion occurs and induces the partial burning of the coke. In the process of combustion, carbon monoxide forms, which reduces the iron oxide ore to metallic iron in liquid form. The liquid iron can then be cast into "pigs" or billets for further processing into steel products.

Because metallurgical-grade coal is one of the primary resource inputs in the making of steel, the demand for it is determined largely by the demand for steel. There are other uses for coke and metallurgical coal, such as in foundries and residential heating, but the amount of coal required for such purposes is very small.

During the past decade the coking demand for metallurgical coal has declined dramatically. In 1984 only 43.8 million tons of metallurgical coal were consumed in coke ovens, as compared with 94.1 million in 1973 (see table 40). Between 1973 and 1984, U.S. coking coal consumption declined by 47%, mostly as a consequence of technological changes associated with the production of domestic steel and as a result of increased imports of steel. The decline between 1981 and 1982, which left 1982 coal consumption at only 39% of its 1973 level, was due primarily to the effects of the deep worldwide recession. Not only did the domestic demand for products produced

with steel (automobiles, appliances, and equipment) decline precipitously, but also foreign steel producers with excess capacity
became progressively more aggressive in world markets, including
in the United States. In this country steel production fell from 120
million tons in 1981 to approximately 80 million tons in 1982. Many
permanent steel plant closings have occurred, and a major contraction of the steel industry is under way. It is unlikely that a surge in
the demand for coking coal will develop in the near future in the
United States. The foreign demand for coking coal, on the other hand,
may well strengthen as foreign steel is imported in ever-increasing
quantities.

In the United States, coke is produced by two methods: the
"beehive" process and the slot-oven process. The first uses refractory
kilns, which produce less than 2% of the coke made; slot-oven coke
plants produce the remainder. Of these, 90% are owned by or are
associated with iron and steel producers who use the coke internally
in blast furnaces. The remaining 10% of the slot ovens are called
"merchant plants," whose operators sell the coke in the open market.

The demand for metallurgical coal is influenced by three technical
factors: the amount of coal required for making a ton of coke, the
amount of coke used to produce a ton of pig iron, and the amount
of pig iron necessary to produce a ton of steel.

The ratio of pig iron consumed per unit of steel produced has been
generally declining. Not only has blast furnace efficiency been improved but also a significant portion of steel is now produced from
scrap metals in electric furnaces. Each leads to reduced demand for
pig iron.

The coke-to-pig-iron ratio also has been declining as blast furnaces
have become more efficient and as supplemental fuels have begun
to be used. This decline has been gradual and steady and is likely to
endure.

The amount of coal required to produce a ton of coke is a function of the quality of the coal and of the efficiency of the oven used.
Historically, the coal-to-coke ratio has varied little: it takes approximately 1.4 tons of metallurgical coal to make 1 ton of coke. This
ratio can vary widely, however, depending upon the particular coals
used for blending. Nearly all coke produced in the United States is
the result of some blending of coals. Furnace operators seek coal that
will "cake" (that is, form a porous, solid mass after being heated).
Bituminous coals used in the making of coke are usually classified
by content of volatile matter (material released during pyrolysis): low,
if they contain 14%-22% volatile matter; medium, if 22%-31%; and

high, if volatile matter exceeds 31%. In addition, coking coal must contain little sulfur and ash. As a rule U.S. producers use high-volatile coal as a base and blend it with low-volatile coal to achieve a mix with the needed specifications. Using low-volatile coal alone could lead to damaged oven walls because this coal is highly expansive.

In regard to the second technical factor in metallurgical coal demand, the use of coke to produce pig iron, several processes exist for converting iron ore into metallic ore (pig iron). At this time the most efficient, least costly, and most widely used processor is the blast furnace. The oldest and most widely used type of blast furnace in the United States is the open hearth furnace. An alternative type of furnace is the basic oxygen furnace (BOF), a piece of equipment developed mostly in prewar Austria and perfected during World War II in Germany. The second technique, which is capable of producing metallic iron at half the cost of the open hearth method, has been introduced in the United States by steel manufacturers only gradually. This is one explanation for the failure of the U.S. steel industry to compete effectively in the domestic market and abroad.

There is an alternative method of producing steel, direct reduction, that uses no coking coal at all. Direct reduction produces steel outside of a blast furnace. Today most direct reduction of iron ore takes place in West Germany and Sweden. In the former, low-Btu synthetic gas is used in the process. Direct reduction in Japan, the world's third-largest steel producer, is used mostly on an experimental basis. Because of the state of economic conditions in the early 1980s, and intense worldwide competition, the U.S. steel industry often seems financially incapable of or unwilling to implement this new technology. If direct reduction were undertaken massively in the United States or elsewhere, the demand for metallurgical coal would decline steeply, but the very high investment costs of building direct reduction facilities preclude their introduction on a wide scale in the United States in the near term. Three existing U.S. direct reduction facilities produced only about one-half million tons of steel in 1980.

Nearly the entire U.S. demand for metallurgical coal is satisfied by Appalachian coal, in particular by coal from Pennsylvania, West Virginia, and Eastern Kentucky. As a share of total coal consumed in this nation, the coking coal market has been declining steadily since 1970 (see table 41). In 1984 this market represented only 4.9% of all coal consumed in the United States, while in 1970 the market share was 3.5 times greater.

Recession-riddled 1982 is perhaps a poor year to use in a review of the recent performance of the coke-producing industry. Declining demand for steel, the greater use of steel substitutes, and increased steel imports have all limited strongly the growth of the domestic steel industry. Moreover, the industry's prospects are poor. To become competitive once again, the steel industry needs to substitute capital for labor. In order to do that it needs a long, sustained economic expansion and the continued retirement of marginal, high-cost production facilities. Only then could the industry and with it the demand for coking coal be revived.

The steady decline in coal consumption by coke plants is also reflected in the statistics shown in table 41. Between 1970 and 1984, demand by U.S. and Canadian coke plants declined 56% for Kentucky coal, 46% for district 8 coal, and 55% for Appalachian coal overall. This implies that Kentucky coal mines shipped relatively less coal to coke ovens in 1984 than did competing producers located in district 8 or Appalachia. The explanation for this phenomenon probably lies in the larger number of small mining operations located in Eastern Kentucky than in the larger regions and in their dependence on spot-market sales. The demand for coal traditionally shrinks first in the spot markets and only later in the contract markets.

Table 41 also shows that Appalachia has been the principal supply source for metallurgical coal and that its share of the North American market has been a fairly constant 90% in the most recent years. The data also show the same stability for district 8 output, as well as the sharp decline of Kentucky production in its share of the market.

THE INDUSTRIAL DEMAND FOR COAL

Coal burned in the industrial or manufacturing sectors of the economy is used primarily to generate steam for process uses, for electricity generation, for space heating, or for a combination of the three. The chemical products industry uses the largest amount of coal of any sector; its needs are for fuel to burn in boilers for process use. The stone, glass, cement, and clay products industries, the second largest industry group to use coal, employ the substance almost exclusively for process heating. Two additional sectors that use coal for boilers are the paper and steel industries. Table 42 summarizes 1984 industrial use of coal by sector. Seventy-six percent of the total industrial consumption of coal was accounted for by four sectors, and there appears to have been no change in the relative importance of these sectors over the past ten years. Unlike the electric utilities and steel manufacturers, industrial consumers purchase coal largely on the spot market.

As in other sectors, the 1973 oil crisis and the subsequent oil price shocks severely depressed the industrial demand for energy, including the demand for coal. This was brought about by the general slowdown in economic growth and by successful and ongoing efforts to conserve fuel. Moreover, uncertainty about the future of the economy and about federal policy toward air pollution standards prompted industry to curtail sharply its demand for new boilers, particularly for coal-fired ones whose initial capital costs are relatively high. Boiler sales dropped precipitously after 1973 and have remained at depressed levels as firms have stretched as far as possible the useful life of their existing boilers; it is not unexpected that coal sales to industry have stagnated over the past eleven years. Table 43 shows the steep decline in industrial coal consumption between 1970 and 1977, a drop of 32%, and the subsequent relative stability in the tonnage sold in that market. It appears unlikely that the future demand for industrial coal will rebound. Rather, as boilers wear out completely and are phased out of operation, they will be replaced by alternatively fueled equipment and electricity, or perhaps in the long run by fluidized bed combustors.

Table 43 also shows the amount of district 8 coal shipped to industrial users. The trend since 1970 had been moderately downward until 1982, but it turned up for 1983 and 1984. Whether 1984 represents the start of a new trend is impossible to determine at this point.

Finally, also of interest is the geographic location of the principal consumers of industrial coal. Table 43 ranks the states that buy industrial coal in the eastern United States in the order of tonnage consumed. Not unexpected is the finding that the largest consuming states are located in the East North Central region, followed by the Middle Atlantic, South Atlantic, and East South Central regions. These regions account for 66% of the industrial coal consumed in the nation. The remaining 34% is consumed west of the Mississippi and is mined there as well.

THE RESIDENTIAL AND COMMERCIAL DEMAND FOR COAL

The smallest and, consequently, least important demand category for coal is that of the residential and commercial user. Demand by this sector has declined from 92 million tons in 1952 to a mere 9.1 million tons in 1984, a drop of 90%. It is generally held that, in today's environmentally conscious world, residential and commercial users cannot burn coal in an acceptable manner. Consequently, it is likely that shrinkage of demand in this market, illustrated in table 45, will continue. Purchasing only 1% of total domestic sales of coal, this sector is not terribly important. Homeowners and the few remaining establishments still using coal as a fuel source will continue to replace their equipment as it wears out with less-expensive units burning other fuels.

IV

SUPPLY OF APPALACHIAN COAL

The supply of coal offered for sale by mining firms and their brokers depends on cost, market, economic, and institutional factors that are in constant flux. The economics of some of the variables that influence supply has already been discussed in chapter 2. In this section historic data are used not only to support the economic arguments but also to complete the framework for analysis of the industry.

Approximately 50% of the national output of coal originates in the Appalachian Coal Basin. Over the past ten years this share in total output has declined steadily. In fact, as table 13 shows, Appalachia produced in 1984 only 5% more coal than it did in 1970, while at the same time national output increased by 38%. The entire rise in U.S. output is ascribable to the vast expansion in western mining, which increased nearly eightfold in that time period.

The statistics in table 46 show the tonnage and the percentage share of total U.S. and Appalachian output contributed by each coal-producing state in the region in 1984. The two largest coal-producing states in Appalachia in 1984 were West Virginia and Kentucky, which together produced 58% of Appalachian and 31% of U.S. output—sizable contributions to coal production.

COAL PRODUCTION IN APPALACHIA

Since 1970, only two Appalachian areas—Eastern Kentucky and Alabama—have increased substantially their coal output. Eastern Kentucky has shown a 70% increase, mostly as a result of the

burgeoning number of surface-mining operations (see table 47).
Alabama has increased output by 27%. The remaining four major coal-
producing states in the region recorded no change or declines in output, Ohio having the severest.

All states in the Appalachian basin except Ohio and West Virginia
expanded their surface-mining operations between 1970 and 1983. In
Virginia this expansion was sufficient to offset the decline in
underground operations; in Pennsylvania it was not. In Alabama as
well as in Eastern Kentucky, both surface and underground operations
produced more coal in 1982 than in 1970. In the latter, surface output increased by 51%, and underground production rose by 13% (see
table 48).

The decline in underground mining which followed the passage
of the 1969 Coal Mine Health and Safety Act lasted for nearly the
entire decade of the 1970s. During these years the added costs
associated with meeting the newly legislated safety standards prompted underground-mining operators, where feasible, to switch to surface operations. The data shown in table 48 clearly document the
change. The trend toward more surface mining was unmistakable for
most basin operations.

Altogether, Appalachian coal output was 8% higher in 1983 than
in 1970 (see table 47). Not until 1979 did the basin's total output sur-
pass its 1970 level of 408 million tons; in 1981 output was virtually
the same as in 1970. In short, as a whole, Appalachian coal output
has remained relatively constant over the past fourteen years. The
same cannot be said for output of its component state regions. Production expanded significantly in Eastern Kentucky while contract-
ing in Ohio, Pennsylvania, and West Virginia. Without the substantial expansion in Kentucky, total Appalachian output would have
declined.

The best illustration of the rapid growth in surface mining and
the decline in underground mining is the share each has held in total
production over the past fourteen years. For Appalachia as a whole
surface mining represented one-third of total mining activity in 1970.
By 1978 that share had risen to one-half, only to decline to two-fifths
by 1983, mostly as a consequence of the restrictive provisions of the
federal Surface Mining Control and Reclamation Act passed in 1977.
The most dramatic rise in the relative share of surface mining occurred in 1978. Immediately following the passage of the act, the share
of deep mining dropped by 16% in one year (see table 49). Since 1979
the relative shares of each type of mining in total output have remained remarkably constant. These shares now differ little, for ex-

ample, from their 1975 levels, the first stable year following the world oil boycott.

Table 49 also shows the steady change in output methods used in Pennsylvania and Eastern Kentucky coal mining. Of all the states that make up the Appalachian basin, only Pennsylvania and Kentucky recorded significant increases in surface-mining output and in their share of total output. In fact, nearly the entire increase in Appalachian production arose from expanded surface activities, in particular in Pennsylvania and Eastern Kentucky. Other changes in the ratio of surface to underground mining occurred in Virginia and Alabama, but they were minor. Output from these states is small when compared with that of central and northern Appalachia and does not make up a very significant part of total basin production.

The rapid increase in surface mining in Eastern Kentucky in the early and middle 1970s is explained largely by the ease of entry into the industry. Relatively lax enforcement of the posting of reclamation bonds, mine inspection, and reclamation regulations and the use of the broad-form deed have traditionally favored surface mining in this coal state. It typically has been easier to start or to expand existing surface-mining operations in Kentucky than elsewhere. Consequently, whenever demand expanded in the short run and prices rose, surface operators developed new mines relatively quickly and captured lucrative economic rents. This occurred in the 1970s even though, in comparison with the coal of neighboring states, Eastern Kentucky's coal lay buried under much steeper mountain slopes. Nearly two-thirds of its coal must be extracted from mountains whose slopes exceed 25 degrees. In contrast, 80% of Pennsylvania's surface output can be mined on slopes of less than 15 degrees, and 55% of West Virginia's output originates on slopes of less than 25 degrees. In short, many Eastern Kentucky surface operators were able to offset the higher mining costs created by steeper mountain terrain with the savings derived from lenient state enforcement of regulations and weaker environmental standards.

With the gradual implementation of the federal Surface Mining Control and Reclamation Act, however, this situation has begun to change. National standards for environmental protection have dramatically increased the costs of entering the surface-mining business. Engineering studies; environmental, reclamation, water and wildlife impact plans; coal transportation proposals; and many other documents must now be submitted to the appropriate state bureaus before a permit is issued. The entry costs have multiplied and have made surface mining in Eastern Kentucky at least as expensive as, and perhaps even more expensive than, that activity in neighboring states. It now appears to be less expensive to purchase West Virginia coal than Eastern Kentucky coal of the same quality, particularly if the destination of the coal is closer to the former than to the latter. Recent surface production statistics support this conclusion. Between 1977 and 1982, surface production in West Virginia expanded by 30% while that of Eastern Kentucky contracted by 6%. The golden years of Eastern Kentucky surface mining, when expanding demand and weak enforcement of minimal reclamation standards allowed the extraction of extraordinarily high profits, seem to be coming to an end. In the future, Eastern Kentucky surface operators will have to compete under equal regulatory terms with their counterparts in West Virginia. Transportation cost differentials and efficiency in mining will decide future market shares.

In Pennsylvania, surface miners expanded output dramatically between 1970 and 1978, nearly doubling production. Since then, consistent with the nationwide trend established by the 1977 reclamation act, surface output has retreated somewhat. Until 1978, deep-mining operators saw their markets being usurped by coal from lower-cost surface mining operations. Since then, however, with the exception of temporary spurts, underground production has remained relatively stable. Output in 1983 was just a little higher than in 1978 and represented somewhat more than one-half of the state's total production (see table 49).

In West Virginia the steep decline in deep mining that lasted until 1978 seems to have abated. In 1978 underground production was 44% lower than eight years earlier, but by 1982 it was only 9% lower than in that year. Surface output for the state in 1982 was not much changed from its 1970 level, but it did decline in the soft market of 1983. Of the major Appalachian coal-producing states, West Virginia, Pennsylvania, and Virginia have been the most stable producers overall for the past fourteen years, their outputs declining only 9% and 10%, and increasing 1%, respectively.

As table 47 illustrates, Ohio's total output decreased by more than one-fourth in the past fourteen years. In Ohio, both underground and surface operations output declined equally. Little structural change in the Ohio coal industry appears to have occurred in these years. Surface operations continue, as they have in the past, to produce two-thirds of total output and underground operations one-third. Overall, the state's coal is high in sulfur and ash content and relatively low in heat content. Such coal is not as widely demanded as is its West Virginia and Eastern Kentucky substitute. Even a new Ohio electric utility has recently contracted to receive more-expensive West

Virginia and Eastern Kentucky coal rather than buy less expensive but qualitatively inferior Ohio coal.

For topographical reasons, surface mining has always dominated coal production in Ohio. The terrain is flatter than in neighboring states and reclamation is therefore less costly. In contrast with the coal reserves of Eastern Kentucky and West Virginia, the largest share of Ohio coal reserves contains a relatively high 2% of sulfur. Consequently, even though surface production is somewhat less costly in Ohio, environmental considerations prevent growth in the demand for this coal.

In Virginia, coal production is dominated by underground operations. Not quite four times as much Virginia coal was mined underground in 1982 as on the surface. The modest growth in output of both types of mining in the state is ascribable to the generally high quality of the coal. This coal is found in the southwestern corner of the state and is comparable in quality to the high-grade southeastern Kentucky coal. But Virginia, despite its significant contribution to Appalachian output, has low reserves. This fact is likely to have a detrimental impact on costs and on the state's ability to compete in the future. Increasingly, as deeper and less accessible seams of coal are brought into production in Virginia, producers will experience the pressures of rising mining costs.

Alabama is the smallest of the six principal coal-producing states. Over the years, surface operations have expanded their share of total output; total output has risen 27% in the past fourteen years. Alabama coal is generally produced for in-state consumption. As the regional economy expands, so will, other things equal, output from Alabama's mines. Unfortunately, low reserve levels in the state may cause mining costs to rise in the future.

In summary, coal production in Appalachia has not expanded greatly in the past fourteen years. The 8% increase is attributable to expanded production primarily in Eastern Kentucky and secondarily in Alabama and Virginia. With the exception of the situation in Ohio, surface operations have expanded their share in total output at the expense of underground production. This trend has been most pronounced in Eastern Kentucky and Pennsylvania. West Virginia remains the largest single producer in the basin, followed most closely by Eastern Kentucky.

COAL RESERVES

For most of the 1970s and the early 1980s, the increasing concern with environmental degradation, including acid rain, has focused attention on the low-sulfur, high-heating-value coals of the Appalachian basin. From the standpoint of quality, these low-sulfur coals are the best in the United States and perhaps in the world.

The quantity of coal which can be expected to find its way to market is, of course, a function of the quantity of the resource which lies in the ground. The largest Appalachian basin reserves of high-quality eastern coal are located in West Virginia and Eastern Kentucky and can be mined either underground or from the surface. Pennsylvania and Virginia also have sizable quantities of the same quality of coal, but they are suitable only for underground mining. Ohio's and Alabama's reserves, although of high quality, are quite small (see table 50).

The most generously endowed state with high-quality, clean coal is West Virginia. With 14 billion tons in reserves, it has twice as much coal as either of its nearest rivals, Pennsylvania and Eastern Kentucky.

Although it would be useful to know how much high-quality coal is mined from under steep slopes and where, there are no recent data linking tonnage produced to the angle of slope. Consequently, it is impossible to assess accurately the future relative mining costs of the different regions of the basin. A study commissioned by the U.S. Council on Environmental Quality fifteen years ago showed that in 1971, 62% of the coal mined in Eastern Kentucky came from slopes in excess of 25 degrees, while only 14% of West Virginia's production originated from such steep terrain.[1] If these estimates continue to characterize surface mining in the future, it is reasonable to expect that, other things equal, high-quality Eastern Kentucky coal will be more expensive to mine than competing coal elsewhere in the basin.

Because of the distribution of high-quality reserves, West Virginia and Eastern Kentucky are likely to continue to be major surface producers in the future. Pennsylvania and Virginia are likely to be sources of quality coal from underground mines. Unless they invest in sulfur removal equipment, electric utilities desiring to meet existing or future clean air standards can be expected to rely heavily on West Virginia and Eastern Kentucky coal. This is true particularly for utilities located in the eastern and southeastern United States because transporting east, far beyond the Mississippi River, the only other clean domestically produced coal (that from the western U.S. coal fields) would be prohibitively expensive. Moderate quantities of clean coal can be expected to come from Pennsylvania and Virginia also.

As we saw in table 9, the West holds vast reserves of coal, and such estimates, although tenuous, imply a lengthy mining life for western coal. The mining life of the eastern coal fields may be somewhat shorter because the eastern fields have been mined longer and more intensively than the western.

Intensive mining has occurred in Eastern Kentucky and Virginia in the Appalachian basin; less intensive mining has taken place in West Virginia and Ohio and the underground coal-mining areas of Pennsylvania. If one accepts the assumption that the best and most easily accessible coal is mined first, one can also predict that, in general, long-term mining costs are likely to rise more rapidly in Eastern Kentucky, in Virginia, and in surface-mined areas of Pennsylvania and Tennessee than in the other states. This trend ought to be viewed as long term, assuming that there will be no change in other factors such as the type of coal mined, topography, and seam width in all of the mining regions.

Table 51 shows an estimate of the mining lives of Appalachian coal areas expressed in years. The estimates were prepared by assuming that 1980 production rates would continue until the reserve base is exhausted or until it becomes economically infeasible to mine the coal.

When estimated remaining production years of all coal reserves are considered (see table 51), Eastern Kentucky, Tennessee, and West Virginia do not stand out. Reserves in Ohio and Pennsylvania (underground) are considerably greater. But looking at only total reserves ignores considerations of sulfur content. Because of the current focus on maintaining clean air, it is more relevant to examine the size of the low-sulfur, high-Btu-content reserves, which, being most heavily in demand, are currently being mined most intensively. Unfortunately, the most recent data available linking reserves and sulfur content were published in 1974,[2] and these data are technically no longer up to date. It can be argued, however, that since 1974 no significant changes in the sulfur content of the estimates of Appalachian coal reserves have occurred to invalidate the earlier estimates. In general, new reserve data tend to increase rather than reduce previous estimates so that the expected longevity of mining described in table 51 is, if anything, conservative. The table also shows the many years of production left in Ohio and Pennsylvania underground mining and the abundance of Ohio, Alabama, and West Virginia surface reserves. In comparison with these and the Pennsylvania underground reserves, the Eastern Kentucky endowment seems rather sparse.

An important assumption that underlies the estimates of remaining mine life is that, on average, only 50% of the underground reserves and 80% of the surface reserves demonstrated will be recoverable in the future. These ratios, commonly applied by the U.S. Department of Energy (DOE), may be somewhat conservative; experienced mine operators often extract considerably more coal from their seams than these ratios suggest. However, mining conditions vary greatly from one region to the next, and what may be economically minable in 1980 may not be so in 1984. Consequently, it seems wise to err, if we must, on the side of a conservative rather than a liberal interpretation.

The values shown in table 51 reveal an interesting aspect of Appalachian coal's future. Although enough coal remains in most Appalachian states to satisfy the needs of this and at least one future generation, the dwindling reserves cast a shadow on what otherwise would be an optimistic outlook. With relatively few remaining years of existing reserves, Tennessee, Alabama, and Virginia are likely to experience increasing costs in underground mining and rigorous competition from West Virginia, Ohio, and Pennsylvania mines.

In surface mining, only West Virginia and Ohio seem to possess sufficiently large reserves to maintain stability in mining costs. Alabama's coal is generally considered of poorer quality and therefore may not be directly competitive with the coal mined elsewhere in the region. For the rest of Appalachia, as increasingly less attractive land is pressed into surface mining, upward pressures on mining costs are likely to develop.

The tenuous nature of reserve estimates should be fully appreciated. Ongoing work is beginning to show that U.S. coal reserves are greater than originally thought. If this turns out to be the case, the mining lives obviously would increase. Many factors, such as location, proximity to rail or water transport facilities and to consumption points, roof conditions, water and gas deposits, and future cost and price considerations are omitted. All of these factors may have significant impacts on recoverability and production. Consequently, the values shown in table 51 should serve as comparative benchmark measures rather than as definitive yardsticks.

INDUSTRY CONCENTRATION

One of the most commonly used and best understood measures of industry concentration is the market share held by the leading producers. Unfortunately, no unanimity exists among experts on the numerical values which best measure concentration. It is generally held that if the four largest firms control less than 50% of the market, concentration and the potential for anticompetitive behavior most likely are not a threat to the industry.[3] Another criterion frequently used in assessing the potential for anticompetitive behavior is the degree of freedom of entry into the industry. If this freedom is largely unrestricted, competition is likely to exist. In short, if concentration in a particular industry is low and ease of the entry high, competitive behavior can be expected.

Evidence suggests that concentration in the Appalachian coal industry is low. In 1983, the four largest coal-mining companies produced 73.3 million tons, only 19.4% of the region's total output (see table 52). This percentage share is nearly the same as the share in output of the four largest coal producers in the United States. The eight largest mining companies in Appalachia held 30% of total Appalachian output, considerably below a 50% market share. The eight largest national coal-producing companies (not the same firms as the eight largest regional producers) also produced 30% of total output. This suggests that both the Appalachian and the U.S. coal industries are unconcentrated.

Changes in the U.S. coal industry have made it less concentrated currently than it was in 1970 (see table 53). One would have to examine 1950 data to find less concentration in the industry than today. The statistics shown in table 53 reveal that the highest level of concentration occurred among the four largest companies in approximately 1970. The year 1974 seems to reveal the highest degree of concentration when the largest eight, twelve, or fifteen firms are considered.

The most recent statistics on market shares of the largest producing firms underscore a now-familiar trend. After the price surges of coal in the middle and late 1970s, large numbers of new, small firms entered the industry. Such entry, particularly in the Appalachian basin, was relatively easy in those years. The requirements to meet strict licensing, engineering, and reclamation standards were not completely in force at that time, and entry costs were correspondingly lower than they are today. Consequently, market shares of the largest firms declined. For example, in 1981 the four and eight largest firms held only 19% and 28%, respectively, of national output, considerably less than in 1970 and 1974. This result, of course, is not unexpected; it reflects the observed strength of the coal market during the post-oil-boycott years. The middle 1980s, however, are likely to show a trend reversal. Entry costs into the industry, due to a host of new environmental requirements, are now considerably higher, and the weak coal market, accompanied by sharply lower prices, has led to the exit of many firms, most of which were small. Producers without long-term contractual agreements who have traditionally sold coal in the spot market are experiencing serious financial problems. Many have already ceased operations, and others are on the verge of halting mining activities. Stockpiles are high at most mines, at rail tipples, and at public utilities. Consequently, industry concentration is expected to rise, but certainly not to a level that would cause concern. It is unlikely that competition in the coal industry will be seriously threatened in the foreseeable future.

The existence or absence of economies of scale can also influence the ease of entry into the coal industry and therefore its potential level of concentration. If significant cost savings can be realized from operating at a very large scale, new firms will be discouraged from entering the industry unless they have available at the outset the large amount of investment capital required to establish an efficiently sized firm. Without constructing facilities equal in size to those already operating at efficient scales, they could not take advantage of the economies of production. They would be forced to mine on a smaller scale with correspondingly higher per-ton costs. In short, they would have difficulty competing with established operations, which over time have been built to an efficient size at which they can capture the available economies. In the coal industry, economies of scale are attainable mostly where the terrain is relatively flat and the ratio of overburden to seam volume is low. This is the case in the western United States, where large equipment is used for surface mining at very low per-ton cost. In contrast, Appalachian terrain and its coal deposits preclude the use of large equipment. The efficient operating size is much smaller in the East than in the West. It is often the case that a seam in an Eastern Kentucky mountain hollow can be mined most efficiently with only a bulldozer, front-end loader, and truck, all rented. The capital requirements for entry in such a case would be minimal, dominated largely by the costs of securing the necessary mining permits and bonds. Hence, raising large amounts of investment capital to attain an efficient scale of operation does not appear to be a serious issue in the Appalachian basin. Scale considerations are generally not a barrier to entry in Appalachia.

Concentration is also measured by mine size. Table 54 lists the mines located in the principal Appalachian states whose production exceeded 1 million tons in 1983. Of the fifty-eight mines shown, only twelve engaged exclusively in surface operations. All others mined coal underground. The fifty-eight mines produced 26% of total basin output. In other words, average mine size in the basin is relatively small; just less than three-fourths of the coal sold was mined by operations with annual production rates of less than 1 million tons. The statistics also show that the preponderance of the operations were underground and that none of the twelve large surface operations were located in Virginia or Pennsylvania. There appears to be no concentration by mine size in the basin, and no single mine was so overwhelmingly large as to have produced an unusually high proportion of the output. Twelve of the fifty-five largest mines are owned by Consolidation Coal Company, eight by Peabody Coal Company, three by Drummond Coal Company, and three by MAPCO, Inc.

FIRM SIZE, PRODUCTIVITY, AND DAYS WORKED

In 1983, 2,778 mines produced 368.1 million tons of Appalachian coal. As shown in table 55, this number includes only mines whose annual output exceeded 10,000 tons. Coal produced at very small operations, some of which were probably unlicensed "wildcat" operations, are not included in the total. Also omitted are the states of Maryland and Tennessee because their annual output is quite small relative to that of their neighbors.

Average mine size differs among states. Size is determined by the types of geological formations present in a state, state mining regulations and their enforcement, the size of the market supplied and the method and ease of access to it, and other factors. To fully illustrate the differences in the sizes of mining operations in the six Appalachian states, it is necessary to develop a size distribution of mining firms in each state and in the Appalachian basin. Table 55 shows such a distribution for 1983. The data show that 61% of the output originated from mines whose annual production is 200,000 tons or greater (categories 1 and 2). These firms represent only 15% of the total number of firms mining in the six-state region. Conversely, 70% of the mines (categories 4 and 5, mines that produced between 10,000 and 100,000 tons annually) supplied only 21% of the output. In short, the preponderance of basin output originated in the larger rather than the smaller mines.

The largest number of mines in any one of the six states shown in table 55 are located in Eastern Kentucky. This state region contained 999 underground and surface mines in 1983, 56% more than West Virginia, the state with the second largest number. Ohio and Alabama, with 185 and 98 mines, respectively, contained the smallest numbers of mines. Mines are separated into underground and surface operations and listed with average output in table 56. From these data we can see that all average underground and surface operations were larger in 1983 than twelve years earlier. For underground mining, Ohio showed the largest average output per mine; Eastern Kentucky and Virginia, the smallest. Ohio mines were on average nearly twelve times as large as their Virginia counterparts. Eastern Kentucky's underground mines were of nearly the same average size as Virginia's. The startling disparity among mine sizes in Appalachia has existed since 1971 at least. In underground mining it seems to have widened somewhat in recent years, while in surface mining the disparity has narrowed markedly.

Also interesting is the fact that the greatest increase in underground mine size has occurred in Eastern Kentucky and Ohio. In the latter, annual output of the average mine has more than doubled in the past twelve years.

The increase in mine size in all states is a function of a number of factors. Prominent among these are economies of scale derived from the required installation of safety equipment. The more stringent standards introduced by the 1969 Federal Coal Mine Health and Safety Act have encouraged larger-scale mining, which allows the distribution of the safety costs over a larger number of tons of output. Also, the cost benefits derived from the unit-train or large-volume transport of coal have made larger operations more profitable. Since the implementation of the health and safety act, there have been no significant new factors that have influenced underground mining. Consequently, one would expect underground operations to have adjusted by now to the new regulations and to reflect the approximate optimal scale in the various state regions. For example, it is well known that the geological formations and the institutional environment of Ohio and Alabama are conducive to large-scale underground mining. In Eastern Kentucky, Virginia, and West Virginia, this is not the case. Smaller-scale operations in out-of-the way mountain hollows, mining relatively narrow seams under generally unfavorable mining conditions, still abound, although their number is declining. The relatively large number of small, though not necessarily inefficient, operations keeps average mine size low in the three states.

With the exception of Ohio, the average size of surface mines in Appalachia has increased significantly in the past twelve years. With the lax enforcement of reclamation standards and favorable court rulings on the obligations of surface mine operators, such operators have found it profitable to expand their scale of operations in Eastern Kentucky. The data detail a threefold increase in the size of the average Kentucky surface mine over the twelve years listed in the table. A similar expansion occurred in Virginia. In West Virginia the expansion of the average surface mine proceeded more modestly, and in Ohio average mine size declined somewhat. In Pennsylvania it nearly doubled.

In 1971 the disparity in average surface mine sizes was quite large among the Appalachian states, a reflection of the differences in environmental standards and the degree of their enforcement and in topographical conditions. The federal Surface Mining Control and Reclamation Act of 1977 introduced enforcement of uniform standards and procedures. Uniform enforcement can be expected to result in a reduction of the disparity in average mine sizes. Potential operators of small surface mines will discover prohibitively high entry costs, and existing small-scale miners will find the costs of meeting the new standards very high. Consequently, the average size of mines can be expected to expand until the optimal scale peculiar to the topography has been reached. This trend seems to be already under way in Eastern Kentucky, Virginia, and Alabama. As table 56 shows, since 1976 the size of the average surface mine has more than doubled in each of the three states. In West Virginia and Pennsylvania, size increased by only 87% and 45%, partly because reclamation standards and their enforcement were already more stringent there than elsewhere in the basin. The same is true for Ohio.

A classification of Appalachian mines by size, type of mining, and percentage share in output is presented in table 57. The table shows, for example, that only 2% of the surface mines and 2% of the underground mines in Eastern Kentucky produced more than 500,000 tons annually in 1983. Their shares in the state region's output were 15% and 28%, respectively. In contrast, 67% of the underground Ohio mines fell into the 500,000-tons-per-year category, and they produced 92% of the state's underground output. Clearly Ohio and Alabama mines are relatively large.

At the other end of the spectrum, 57% of the Eastern Kentucky surface mines were in the smallest size category, producing 50,000 tons or less annually. Similar percentages for this category are evident for surface operations located in the other states. Underground operations, on the other hand, showed a similar preponderance of small operations only in Virginia and Eastern Kentucky. Underground operations in other states are considerably larger.

The comparative size distribution of Appalachian mines and their share in output are diagramed in figures 4 and 5, which illustrate the relationship between the cumulative number of mines and their share in surface and underground output. Only mines with an annual out-

Figure 4. Relationship between Percentage of Underground-
Mining Operations and Percentage of Coal Produced, 1983

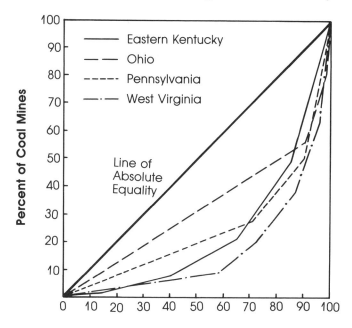

Percent of Output from Underground Mines

Source: Table 57

put of at least 10,000 tons are shown for Eastern Kentucky, Ohio,
Pennsylvania, and West Virginia.

The 45-degree line of absolute equality in figures 4 and 5 denotes
the situation in which a given percentage of mines contributes an
equal percentage of output. This equal distribution would exist, for
example, if 16% of the mines in a state produced 16% of total output. The more concave (from above) the curves are, the greater the
share of total output accounted for by a small percentage of mines.
Figure 4 clearly shows that for the four states, a relatively small
percentage of the underground mines accounted for a disproportionate
share of output. This was most noticeably the case for West Virginia,
where in 1981 6.8% of the mines produced over 40% of the coal mined
underground. At the other end of the spectrum is Ohio, in which 60%
of the mines produced 86% of all output and 80% of the mines produced 96% of output. In short, figure 4 shows that in underground

Figure 5. Relationship between Percentage of Surface-Mining
Operations and Percentage of Coal Produced, 1983

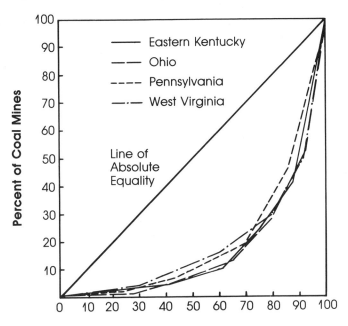

Percent of Output from Surface Mines

Source: Table 57

mining a relatively small number of Eastern Kentucky and West
Virginia mines contributed a major share of underground output. The
contribution of each mine is relatively smaller in Pennsylvania, and
in Ohio the distribution of output is more equal among producing
mines. Hence, the likelihood of market domination by a few firms
is lower in Ohio and Pennsylvania than in Eastern Kentucky and West
Virginia, even though the latter two contain a very large number of
coal-mining firms. Their numbers are large in Kentucky and West
Virginia, but their average output is relatively small.

Figure 5 illustrates the relationship between the number of producing mines and their shares of surface output for the four states
examined. In all states, a small percentage of the mines accounted
for a disproportionately large share of total surface production in 1981.
Mines whose annual output exceeded 200,000 tons (categories 1 and
2) represented 13%–20% of the total number of surface mines

operating, but their share in output ranged between 50% and 64%. At the other extreme, the smallest category of mines, those whose output lay between 10,000 and 49,999 tons per year, represented a remarkably uniform 19.6%–22% of the number of mines operating and produced between 11% and 14% of the output. The similarity among surface mine contributions in the Appalachian states is notable because topography, including steepness of slope and depth of over-burden, varies significantly from one state to the next. In the steepest terrain, found in Eastern Kentucky, a considerable amount of coal is mined from mountains whose slopes exceed 25 degrees; the flat-test terrain is in Ohio. Despite these differences, the percentage shares in output from each size category among the states are remarkably similar. This similarity was much greater in 1981 than in 1973,[4] a finding that suggests that average mine size and the share distribu-tion of different size groupings is approaching uniformity throughout Appalachia. One explanation for this tendency is the trend toward uniform basinwide enforcement of land reclamation standards.

The most widely used measure of efficiency in any economic sec-tor is productivity, which is expressed as the amount of output pro-duced per unit of input, or per combination of inputs. Because out-put for inputs such as capital is difficult to measure uniquely, a con-vention has developed to measure productivity mainly in terms of output per unit of labor input. Economists recognize, however, that changes in labor efficiency are not alone responsible for changes in productivity. Productivity also depends on the type and the quantity of capital equipment and of materials available to labor. In the coal industry it also depends on topographical and geological formations. Over time, attempts have been made to measure output in terms of capital or machinery units of input, but because of the special em-phasis placed on human capital and the relative ease of measurement, the focus remains on days of labor.

For nearly the entire decade of the 1960s, productivity in Ap-palachian underground mining was on the rise. Pressures from abun-dant oil, competition from surface mining, and weak coal markets in general provided underground operators with strong incentives to improve efficiency.

Until the middle 1960s, Eastern Kentucky lagged behind other regions in mining productivity. This was due mostly to the fact that deep mining, then the dominant form of extracting coal, was con-ducted in relatively small mines with inadequate equipment. This is no longer the case. In 1982 Eastern Kentucky labor ranked as the most productive in the four major coal-producing states in Appalachia.

Figure 6. Productivity in Underground Coal Mining for Four
Appalachian Coal Basin States, 1972-1983

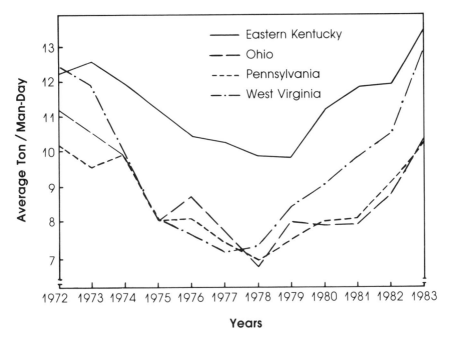

Source: Table 61

Table 58 illustrates the relative productivities in underground and surface mining in 1982 and 1983. In underground operations Eastern Kentucky was 34% more productive than Ohio and Pennsylvania; in surface operations, 10% more productive than Pennsylvania. Kentucky workers also were more productive than the average Appalachian mine worker in both types of mining.

The enactment of the Federal Coal Mine Health and Safety Act of 1969 had, as expected, an adverse impact on underground productivity in all four regions, but by 1978 the impact had run its course. Aided by strong demand for coal after the doubling of oil prices, productivity began to recover vigorously. As figure 6 illustrates, this trend seemed to be continuing into 1983. So far, the economic recovery of 1983 has aided the underground-mining industry in reattaining approximately its 1973-1974 level of productivity.

Two factors, one negative and one positive, explain most accurately the change in underground productivity. The new safety regulations altered work procedures underground and on the surface. A greater share of a miner's workday was required to be spent on safety-related procedures. New personnel were employed to compensate for lost production time at the coal face and elsewhere and to monitor safety conditions. The net impact was to increase the total number of labor hours required to mine a ton of coal; hence productivity, or output per worker hour, declined.[5]

A second factor moderated the decline in average productivity. Passage of new safety standards forced underground operators to acquire new safety equipment. Additional safety training also became a requirement. Many operators who had been only marginally profitable in the past were unable to meet the new costs and left the industry. These were typically the smallest, least productive operators whose output costs exceeded the industry average. Their departure had the effect of gradually raising average productivity for the industry. Consequently, overall productivity in underground mining fell less than it would have if marginal operators had continued to operate.

Eastern Kentucky contained larger numbers of these marginal operations than any other part of Appalachia. Their departure from coal mining made the resulting decline in average underground productivity less severe in Eastern Kentucky than in other Appalachian regions (see figure 6). In 1983, underground operations remained substantially more productive in Eastern Kentucky than elsewhere in the Appalachian basin.

Productivity in surface mining declined substantially between 1972 and 1977 (see figure 7). Starting from an average of more than 30 tons per worker day in 1972 (three times the level of productivity in underground mining), productivity declined steadily in Eastern Kentucky, Ohio, and West Virginia until 1977. In Pennsylvania productivity showed a more erratic pattern, reaching in 1977 a level approximate to that in Eastern Kentucky and Ohio.

The general decline in surface-mining productivity is due to several factors. First, the new, more stringent safety standards imposed on deep mining made surface mining a relatively more attractive alternative. In addition, a slowdown in publicly funded construction of roads, dams, and bridges induced general contractors to search for other opportunities that would allow them to use their idle equipment. Most, however, were inexperienced in surface mining, and their levels of productivity were correspondingly low. This influx of new operators kept industrywide productivity lower than it otherwise would have been. Also, the early 1970s witnessed widespread con-

Figure 7. Productivity in Surface Mining for Four Appalachian
Coal Basin States, 1972-1983

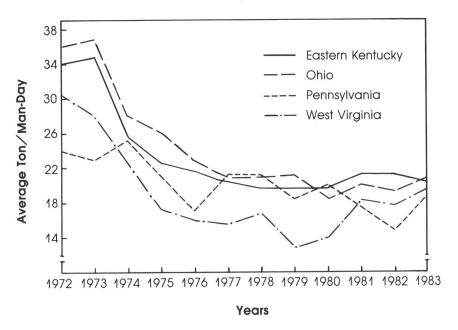

Source: Table 61

cern with environmental degradation, including the disturbance of
environmental balances. This concern culminated in new state and
federal regulations. Operators were now required to restore mined
land more carefully and without long-term damage to the environment. More labor time was used to prevent landslides, sedimentation, erosion, and acid pollution. Not unexpectedly, productivity
declined. The decline was steepest where enforcement and standards
had been weakest: in Eastern Kentucky and Ohio. By 1977, the
disparities in productivity levels among at least three of the four major Appalachian states had been virtually eliminated. Only West
Virginia still showed considerably lower levels than its neighbors.
By 1981 even this state region had improved its performance in the
surface sector, and productivity in all four major Appalachian states
had become remarkably uniform.

Differences in reclamation requirements and in their enforcement
are not the sole explanation for differences in productivity. Geological
and topographical formations of minable land also have a major im-

pact. One would expect Ohio surface operators to be more productive than their counterparts elsewhere in the basin because Ohio terrain is considerably gentler than that of the other states. Even so, since 1977 productivity patterns have been remarkably uniform in Ohio, Eastern Kentucky, and Pennsylvania. Obviously other variables such as height of overburden, seam thickness, ease of access, and rate of exhaustion of reserves also influence productivity and cost. A large body of literature that deals with the theory of exhaustible resources asserts a priori that marginal productivity declines (hence marginal costs rise) with the cumulative extraction of coal. High-productivity, low-cost deposits are mined first and are progressively depleted.[6] Other explanations, however, challenge the notion of high-productivity, least-cost production first and point to the existence of a least-cost-production-last pattern.[7] Further exploration of the impact of many of these influences on productivity and cost would contribute greatly to an understanding of mining practices.

There exists some evidence that the most attractive strippable reserves have already been recovered, particularly in West Virginia and Eastern Kentucky.[8] Consequently, unless technology changes, future gains in productivity in surface mining are likely to be modest. This view may be at least in part an explanation of the relative constancy of productivity ratios in Appalachia observed over the past six years.

A question frequently raised in the literature concerns the relationship—if any—between mine size and productivity. To what extent does the size of the mine, measured in terms of average annual output, influence productivity? In an effort to gain insight into this question, coal output per day of labor was regressed on average annual production per mine for a range of years for the major coal-producing states in Appalachia. No controls were introduced for differences in institutional environments—such as differential safety requirements, enforcement, and topography—because data were unavailable. The past twenty-two years were divided into two time periods, 1960-1969 and 1970-1982.

The results for the first time period show that 52% of the systematic variation in Eastern Kentucky, Pennsylvania, and West Virginia surface-mining productivity is explained by differences in average mine size. Figure 8 shows this relationship graphically. As the size of surface operations expanded, productivity rose.[9]

A similar relationship was observed for Ohio surface mining. Here 71% of the variation in productivity is explained by mine size, but the size of the sample is small (see figure 9). Ohio is separated from

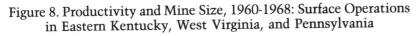

Figure 8. Productivity and Mine Size, 1960-1968: Surface Operations in Eastern Kentucky, West Virginia, and Pennsylvania

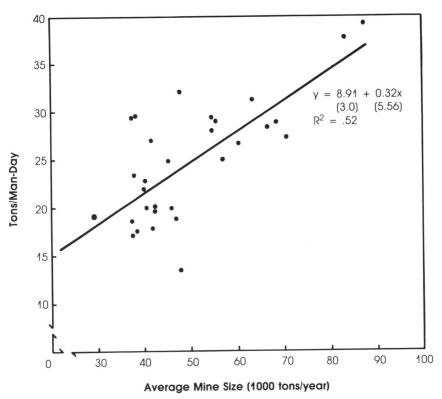

$$y = 8.91 + 0.32x$$
$$(3.0) \quad (5.56)$$
$$R^2 = .52$$

Tons/Man-Day

Average Mine Size (1000 tons/year)

the rest of the Appalachian states because topography and geological formations differ there.

In underground mining, the explanatory power of mine size for productivity is weaker (see figure 10). For the 1960-1969 period, only 40% of the variation in productivity in Eastern Kentucky, West Virginia, Pennsylvania, and Ohio is explained by variations in mine size. These results ought to be interpreted with caution, however, because the output variable also shows up in the dependent productivity measure.

For the period 1970-1982, the regression results were disappointing. The data did not statistically link variations in productivity and in mine size. Too many factors other than mine size seem to have influenced output per laborer day. New safety legislation, new en-

Figure 9. Productivity and Mine Size, 1960-1969:
Surface Operations in Ohio

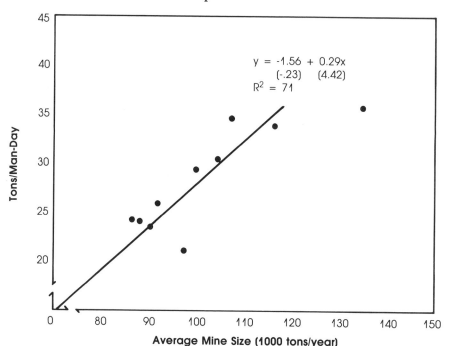

vironmental and land reclamation standards, and the oil boycott and the subsequent surge in energy resource prices combined to make a statistical analysis of the variations in productivity extremely difficult, if not impossible.

A statistic commonly used to assess the stability of a state's coal industry is the number of days mines operate in a year. A high and stable utilization of existing mining facilities over a number of years usually means a stable economic environment with a stable work force. Frequent fluctuations in utilization imply the opposite.

Table 59 shows the average number of days that underground and surface mines were in operation during 1970-1983. For underground operations, the first five years of the 1970s were good, largely because of the oil boycott and spiraling oil prices. In 1975 mines were operating at capacity, satisfying strong demand. Days worked exceeded 200 by a considerable margin. The United Mine Workers of America (UMWA) strike that ran from December 6, 1977, to March 25, 1978, led, however, to a substantial reduction in days worked.

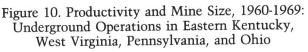

Figure 10. Productivity and Mine Size, 1960-1969:
Underground Operations in Eastern Kentucky,
West Virginia, Pennsylvania, and Ohio

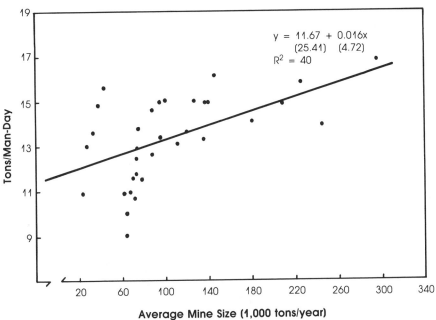

Not unexpectedly, this reduction was most dramatic in the most unionized coal state, West Virginia, where days worked declined to 131, a 41% reduction from 1975. None of the other three states listed in table 59 recorded as severe a decline as West Virginia; their reductions ranged from 25% to 30%.

Underground operations recovered substantially in 1979 and remained at high levels until 1981, when a seventy-two-day UMWA strike once again closed many mines. Recovery from the strike started in 1981 and proceeded through 1982. Because of the worldwide recession the improvement was weak, and mine days for 1982 changed little from their 1981 level. In 1983, mine days continued at relatively low levels.

Among the four states shown in table 59, the 1981 UMWA strike caused the smallest decline in the number of mine days worked in Eastern Kentucky, and the largest, in Ohio. In Eastern Kentucky small nonunion operations abound, and their workers do not necessarily cease to operate when fellow miners at other companies go on strike. In fact, in an effort to supply a strong strike-induced spot-market demand, nonunion miners often work harder and more days. In a sense they represent a buffer of supply sources that tend to moderate escalating prices. They continue to supply coal to the spot market, thereby keeping prices from spiraling out of control. They also tend to mitigate union influence in the industry. It has been noted that the most productive underground-mining operations (see table 58) are located in Eastern Kentucky and West Virginia, and that these are on average smaller than their counterparts in Pennsylvania and Ohio (see table 56); they are also in operation fewer days per year. One must conclude, by inference, that economies of scale, at least for deep coal deposits in Eastern Kentucky and West Virginia, may be realizable at relatively modest rates of output. The large-scale operations found in Pennsylvania and Ohio are probably appropriate for Kentucky and West Virginia.

The number of days worked by surface-mining companies in the four states has risen steadily in the past five years. It is somewhat surprising that with the exception of the Eastern Kentucky area, surface operations in the Appalachian states were mining coal more days each year than their underground counterparts. One would expect surface operations to be more sensitive to weather conditions than underground operations and hence to show fewer average working days per year, but the contrary seems to have been the case. Only in Eastern Kentucky were days worked by surface operators fewer or the same as days worked by underground miners. The one Kentucky exception came in 1981, when the reduction in output from the strike-closed underground mines was offset by vigorous surface activity.

In West Virginia, Pennsylvania, and Ohio also, the difference between output levels was particularly great during the 1981 strike year. Operating days dropped significantly for Pennsylvania and Ohio underground mines and only modestly for surface operations. In West Virginia, on the other hand, surface-mining days dropped a substantial 14%. The contention can be made that, unlike operators in the other states, West Virginia surface operators are sensitive to strikes in the underground sector.

Stability of production levels seems to be highest in Eastern Kentucky underground mining, which shows lowest fluctuations in mine days worked in the past five years. Ohio and West Virginia data show the greatest fluctuations; Pennsylvania's levels are between. Underground mining, output, employment, and income have been relatively more stable in Eastern Kentucky over the past five years than in the other three major states. However, this stability is at least partially offset by the fact that Eastern Kentucky and West Virginia mines operate on average 12% fewer days per annum than Pennsylvania and Ohio mines.

Pennsylvania surface operations recorded the fewest fluctuations in mine days worked over the past five years, and West Virginia recorded the most. Pennsylvania and Ohio surface operations are active nearly 25% more days per year than West Virginia and Eastern Kentucky operations, and they are relatively more stable; that is, fluctuations from one year to the next are lower in magnitude. One explanation may be that a larger share of the Pennsylvania and Ohio surface output is captive (controlled by a parent company) than of the Eastern Kentucky and West Virginia output. Captive mines are less sensitive to the vagaries of the coal market and less subject to the pressures of market price on output and profitability, Also, captive mines often are subsidized by parent organizations for long periods of time even though short-term price considerations would dictate temporary closures because exit and entry costs would otherwise be too high.

THE PRICE OF COAL

The price of coal is an important barometer of the industry's welfare. In the producing states ebullience reigns when price rises, and gloom sets in when it falls. Many producers regard a world of demand, supply, and their influences as too difficult to comprehend and choose price as an attractive proxy indicator for the state of their world.

From the standpoint of economic efficiency, the ideal price for a commodity traded in a market is the equilibrium price.[10] If, for example, the equilibrium price of coal is attained, the market will be cleared of the mineral, and all suppliers and demanders of coal will go home satisfied; no unwanted shortage or surplus will exist. Unfortunately, equilibrium is an elusive target toward which we may expect market price and quantity to gravitate, without necessarily reaching it. In a dynamic market, circumstances that influence demand and supply are in flux. Sometimes these circumstances are mutually supportive of each other; at other times they are at odds with one another. For example, a strike by miners would cause a market price and quantity to move away from equilibrium by reducing the amount of coal offered for sale. An interruption in the flow of oil to market would reduce its availability and precipitate an increase in the demand for coal. If these two events occur at the same time, the resulting pressure will cause an increase in the market price.

Only if all the forces that influence supply and demand are assumed not to change can it be said that the market will clear at some price and that equilibrium will prevail.

Under unchanging conditions, a decline in price will generate an increase in the quantity demanded (consumers want more of a good if its price is lower) but also a decrease in the quantity supplied (suppliers are not so interested in sending the good to market for the lower price). If the decline in price starts from a level above equilibrium (known as a level of surplus), then price will fall until it reaches equilibrium. If a price is below equilibrium, quantity demand exceeds quantity supplied, and a shortage prevails. Here we expect to see the price rise to equilibrium as eager demanders snatch up the bargain and make obvious their demand for more; suppliers will respond by putting increased amounts of the good on the market.

The magnitudes of the required changes in price and quantity needed to attain equilibrium can be estimated only if statistical information is available on the sensitivities of quantities supplied and demanded to changes in price. This sensitivity is a function of time and a host of other factors such as income levels, the prices of competing and complementary products, population, the costs of production, technology, and institutional characteristics. These factors influence the quantity of coal demanded and supplied not only at a particular point in time but also over the long run.

Two markets exist for coal. They differ on the basis of the amount of time agreed upon between buyer and seller. The spot market has a short-term connotation; in it, sales of currently supplied quantities take place. In the contract market, which has a long-term connotation, sales are of obligations to supply a quantity of coal over a specified time period.

No statistics are available that show price and quantity movements for the two markets separately, but delivered prices of purchases of spot and contract coal by electric utilities throughout the nation are published by the U.S. Department of Energy, Energy Information Administration. These statistics, shown in figure 11, are of limited use because they include not only the mine-mouth price of coal but also transportation costs. Spot prices reflect a much higher volatility than contract prices, with some time lag. The sharp drop in 1983 of spot prices to a level below the contract price is interesting. This decline presages falling contract prices for at least 1984 and probably 1985. For the first time in the past ten years, spot prices have fallen below contract levels. Surges and declines in the demand for coal are clearly reflected in the peaks and troughs of the DOE spot-price trend line; much greater stability is reflected in the contract-coal-price trend line.

Figure 11. Coal Prices Paid by Electric Utilities on Delivery,
by Type of Purchase, 1973-1983 [in current dollars]

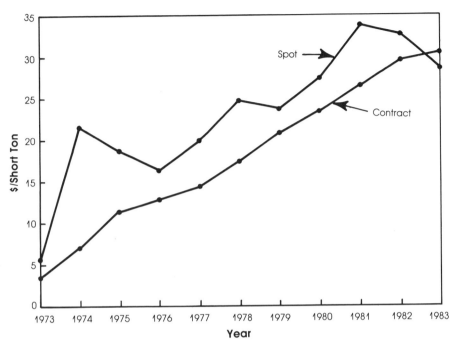

Source: Weekly Coal Production, DOE, EIA, January 27, 1984

Sound theoretical reasons explain why the price of coal sold on
the spot market fluctuates a great deal more than does the contract
price. The former price usually reflects with considerable accuracy
not only immediate market tightness or softness, but also speculative
expectations and behavior. These fluctuations, which may be of considerable magnitude at times, are only gradually incorporated into
the prices of newly negotiated contracts. If the fluctuations in spot
prices reflect only short-lived market disturbances, they may not be
mirrored in contract prices at all. If the disturbances are of longer dura-
tion, contract prices will reflect them eventually.[11]

For a market disturbance to be reflected in the spot-market price,
it must also be unexpected—an interruption in the flow of oil, an
unusually long labor strike, or severe climatic conditions, for example. Disturbances that can be anticipated, such as general price inflation or new or increased taxes on coal, already provided for in most
contractual agreements,[12] would almost certainly be expected to affect the spot-market price in a predictable, moderate volume and direc-

tion. In short, only severe, unanticipated disturbances lead to the sharp price changes observed from time to time in the spot market.

The amount of coal sold in the spot market is relatively small. Estimates put it at no more than 20% of total coal sales, and it is probably considerably less.[13] Consequently, published data on average coal prices generally reflect contract price levels. Most coal is sold on contract because both buyers and sellers prefer stability. Buyers above all seem to value a secure and reliable source of supply, even if they have to pay above-spot-market prices. Having assurance of quality and reliable delivery is well worth the additional cents per ton of coal to electric utility purchasers. Spot-market transactions are used by such buyers mostly to attain flexibility in building up or drawing down inventories.

Multiple-year, long-term contracts give suppliers the ability to secure favorable financing, purchase the necessary capital equipment, and bargain for attractive rail transport rates. In addition, a long-term contract gives a mine operator the security to participate in the spot market. The contract assures him of a base income for his operations and provides incentive and the ability to expand production when price rises and curtail it when price falls. Such producers' activities in the spot market tend to moderate price swings.

Contracts are not inviolable, however. At times sharply rising spot prices have lured suppliers to abandon their contractual obligations and seek irresistible windfall profits obtainable on the spot market. Conversely, in times of persistently declining demand for electric power, electric utilities have broken their contracts and been unwilling to accept all of the coal they obligated themselves to buy. In both cases, litigation usually produces a compromise.

In late 1973 the cut in Arab oil production, accompanied by the embargo on oil shipments to the United States, dramatically increased oil prices. Fears of severe supply interruption caused a sharp increase in the demand for coal in 1974, even though most utility oil imported to the United States came from non-Arab countries. Labor unrest and anticipation of a long coal strike prompted utilities to build up sharply their coal inventories in preparation for the winter months, exacerbating an already tight coal supply situation. As is illustrated by the statistics in table 60, by 1975 the 1973 price of Appalachian coal had more than doubled. The sharpest rise occurred in the price of coal mined underground in Virginia, Eastern Kentucky, Pennsylvania, and West Virginia, the higher-quality Appalachian coals. The lower-quality Ohio and Alabama coals recorded slightly smaller price increases. After that year, prices for nearly all Appalachian coal mined underground began to decline.[14] By 1983, all prices were lower in real terms than in 1975 or 1978. Only Ohio coal prices remained relatively constant.

The lowest 1982 per-ton price average for underground coal, $15.20, was recorded in Eastern Kentucky; this price represented a 33% real decline from the average 1975 Kentucky price. The highest 1982 price was obtained for Pennsylvania coal: $18.49 per ton. Alabama and Ohio prices are considerably higher yet, even though these coals are of lesser quality. This unusual situation is perhaps the result of particularly favorable contractual agreements in force in these two states.

That price competition is present in the Appalachian coal market is reflected in the similarity of the 1983-to-1973 price ratios of the Appalachian states. Table 60 shows that 1983 real prices for underground coal were 1.28-1.54 times their 1973 levels, a relatively uniform change across five states. (Ohio is an exception at 2.60; its low starting price and fortunate contractual situation explain the unusually high ratio of change.) This implies that the differences in 1973 price levels among the Appalachian states remained largely unchanged in the succeeding nine years. Events in the energy resource markets affected the coal industries of the various Appalachian states equally; otherwise, relative prices among them would have changed.

A similar picture emerges for surface-mined coal. In the Appalachian basin, real prices approximately doubled between 1973 and 1983. In Alabama, average price rose to 240% of its previous level.

The rate at which the price increased was not uniform; Eastern Kentucky's coal price in 1975 was 82% higher than the 1973 price, but Eastern Kentucky operators were able to expand output quickly, thus mitigating the potential impact on price of a rapidly escalating demand. Virginia coal, on the other hand, was selling in 1975 for 2.5 times its 1973 price. Most other Appalachian states also were unable to satisfy rapidly the surge in demand. Consequently, prices in these states jumped more steeply, particularly in West Virginia and Virginia. As output adjusted to higher demand levels, and because demand itself was not maintained, prices soon began to fall sharply. By 1983, real prices for surface-mined Virginia and West Virginia coal had dropped by 27% and 28%, respectively. The rapid adjustment of production in Eastern Kentucky caused real prices in the state to rise gradually and over a longer time period, until 1978. Other states' prices reached highs around 1975; by 1978 these prices had already begun to decline.

The 1983-to-1973 ratios for surface-mined coal from the basin states show the same type of similarity as the underground coal ratios showed. Prices generally doubled for coal from all states. (Alabama provides the exception.) Apparently in both types of mining some price differences among states are likely to persist, but competition keeps those differences within acceptable bounds and forces price toward long-run equilibrium.

Any discussion of price would be incomplete without mention of the link between price and productivity. The linking influence between the two is production costs. Higher labor productivity (as long as all other factors remain unchanged) means lower per-unit production costs; as costs fall, suppliers can charge a lower price and still make a normal profit from the sale of their product. Productivity decreases mean higher costs and higher market prices. Prices may not always rise or fall absolutely in response to changes in productivity; rather, changes in productivity can be expected either to moderate or to exacerbate existing price trends.

Productivity in underground mining declined until approximately 1978, after which it began a gradual increase. By 1982 it had recovered roughly to 1973 levels (see figure 6). Spurred by the 1973-1974 world oil crisis and its aftermath, the price of coal surged upward. Following market readjustments, however, price began a steady decline that continues to the present. The effects of this decline on profitability of mining firms was at least in part offset by the rise in productivity and its impact on production costs. Unfortunately, it is impossible to estimate empirically what share of the decline in real prices is attributable to rising productivity.

The link between productivity and price is also apparent if one considers the capacity of an operator to mine coal. The relatively brief 1973-1974 energy crisis caused widespread alarm among the electric utilities, and they began to bid up the price of coal rapidly. In the Appalachian coal-producing region most suppliers were unable to adjust quickly to supply coal when the utilities demanded it, even though they might have wished to. Appalachian producers did, of course, see the opportunity to realize high (economic) profits if they were able to produce much larger quantities of coal very quickly. In markets where supply remained relatively unchanged in the face of increased demand, buyers found it necessary to bid the price of coal higher and higher to be able to obtain the product. This inflexible supply situation existed in most areas of the Appalachian region.

It did not exist in Eastern Kentucky. There, capacity to mine was more flexible. Many small operations were initiated quickly by families and small independent producers who recognized the opportunity for profit. Existing operators expanded their working hours and

operations. Eastern Kentucky had the labor force and an accom-
modating geological structure to enable a surge in output. Buyers in
the Eastern Kentucky market did not have to bid price up as high
as in the other Appalachian regions to obtain coal. Price rose less
rapidly than in the other markets precisely because Eastern Kentucky
mines summoned the required inputs to meet the demand for their
product.

Table 61 illustrates that in 1983 the real prices of coal mined
underground reflected the varying levels of productivity in the major coal-producing states in the basin. Eastern Kentucky producers,
the most productive in the basin, sold coal at the lowest prices. Ohio
producers had the least-productive mines and obtained the highest
prices.

Surface operations show a weaker correlation between productivity and price. In general, surface productivity is considerably higher
than underground productivity, and the prices of surface-mined coal
are lower. But among the states listed, those with the highest levels
of productivity do not necessarily sell their coal at correspondingly
low prices. In 1983, Ohio surface operators show slightly higher levels
of productivity than their Eastern Kentucky counterparts, and the
average real price of their coal is lower. But other factors such as quality and rank of coal, proximity to consuming utilities, and long-
standing relationships between seller and buyer also affect price.
Similarly, Pennsylvania surface operators, the least productive of
operators in the four states shown and 20% less productive than
Eastern Kentucky operators, sell their coal at only slightly higher
prices. One would infer that Pennsylvania surface operations are less
profitable than the Eastern Kentucky ones.

Just as the section on productivity concluded with a word about
captive mines, a discussion of parent-subsidiary relationships is essen-
tial here. Information is published on captive prices (the prices of coal
sold to parent companies by their captives, usually wholly owned subsidiaries). Table 62 shows for the major coal states in Appalachia the
share of total output produced for sale on spot and contract markets
and the share produced for parent-company consumption. A considerable portion of the coal output from Ohio and Pennsylvania never enters the open market because of the concentration of steel-
producing firms in these states. To a lesser extent this is also true
of output from West Virginia, where many mines are owned by steel-
producing firms located in Ohio and Pennsylvania. Eastern Kentucky,
on the other hand, sends little of its coal to parent companies.

Accompanying the decline in steel production has been a decline in the percentage of coal shipped by captive mines. In Pennsylvania, their share in total output declined from 31% in 1973 to only 21% in 1983. For Eastern Kentucky, little change occurred over that period; for the remaining two states no particular trend is evident.

Prices of coal sold by captive mines are generally significantly higher than those observable in the open market. In the first place, captive coal is high-quality coal frequently used in the making of coke. Its superior quality commands a higher price, even when sold in a captive market. Second, the transaction price for captive coal is not freely established and has not passed the market test. It may be an accounting price used by parent firms as an input in the calculation of steel production costs, or the firm's objective may be to minimize tax liabilities in various states. In short, the captive price is an unreliable statistic that reveals little about the behavior of demand and supply parameters in the market.

V

REGULATION IN THE COAL INDUSTRY

Twenty years ago U.S. industry faced few environmental limitations. The environment was regarded largely as a public resource for all to use at no cost. As the demand for environmental use expanded, however, scarcities soon developed. Today the environment is an important consideration in nearly all economic activities. It has become clear that at zero cost too many of the good qualities of the environment will be consumed to the detriment of public welfare. The quality of air, land, and water resources, together with general health and safety, has deteriorated at an unacceptable rate; environmental resources have been depleted much more rapidly than they can be replaced.

Under most conditions, markets can efficiently solve the problem of resource allocation. Prices, fluctuating in accordance with aggregate levels of demand for and supply of goods, represent signals to buyers and sellers on which they base their behavior. Those who respond properly survive and profit; those who do not, fail. In the process, resources are allocated optimally. In a simple abstract model of an ideal economy, the system may well function as outlined. In the world around us, however, relationships are much more complex, and the market system often fails.

From the standpoint of resource management, which is at the core of most regulation of the coal industry, the market has failed often. Two major sources of market failure can be distinguished, both of which result from the simple fact that everyone's economic activities directly or indirectly affect the well-being of others. Many of these effects are termed *externalities*, and they are ubiquitous.

First, environmental resources are not governed by a well-defined and enforceable set of property rights. These resources are not owned privately and therefore are not sold in the market. Because the absence of prices precludes their sale, no market signals exist concerning their use. Second, the public-good nature of many environmental resources requires some form of nonprivate (that is, public) intervention in the market. The intervenor is usually the government or a quasi-public institution, and its role is to ascertain that an appropriate quantity of environmental resources is used or preserved by society.

EXTERNALITIES AND THE MARKET

Economics teaches that the consumer should be willing to pay a price equal to the total marginal social cost of producing the commodity he or she uses. There seems to be no compelling reason to abandon this principle when, in addition to the private costs, external costs of production arise; hence, the total marginal social cost should include both private and external components if the latter are greater than zero. External costs arise whenever the cost of producing a commodity, as viewed from society's standpoint, diverges from the cost of producing it as viewed from the individual's standpoint. Unfortunately, external costs are often ignored by the market, and the true costs of production, therefore, are understated. As a result society consumes too much of these underpriced goods and too little of other goods, and resources are allocated suboptimally. The competitive price system is based on the premise of full-cost pricing, that is, that the total cost of producing a commodity is borne by the person using it. If the user bears only part of the cost, the private cost, resources are not put to their socially optimal use, and someone other than the user bears a portion of the cost.

Consider, for example, the environmental impacts of surface mining that led to the passage of the federal Surface Mining Control and Reclamation Act of 1977. Creeks, rivers, roads, and public forests have ill-defined property rights, but surface mining has clear third-party effects. The removal and discharge of mining residuals can adversely affect the well-being of third parties because neither the mining operator nor the purchaser of coal has an economic incentive to reimburse third parties for their loss in welfare. Even if operators or purchasers wanted to, no impartial mechanism exists to establish a price, and others are forced to bear the adverse external costs that can arise from mining activity. Consequently there have developed, in recent years, many public authorities designed to internalize these third-party effects so that they are borne, in the long run, by the ultimate user of coal—the consumer of electric power or, in the case of metallurgical-grade coal, the users of steel. The failure of the market to allocate environmental resources efficiently has led to its replacement by government institutions.

Another dimension of market failure in allocating some resources involves the public nature of certain commodities and services. Because some goods are distributed without charge, it is not feasible to collect revenues from those who consume them. It is no more possible to sell national defense protection than it is to sell the national prestige of a president or of an Olympic team. It is true that some quasi-public goods are sold, at least partially, such as access to (state university) education or to public parks. But the revenues collected rarely defray expenses, so that ultimately the use of public funding becomes necessary. It is also not possible to exclude people from the use of true public goods, such as clean air or unpolluted lakes.

Whenever a good or a service has a publicness about it, its full marginal cost cannot be captured easily or completely. Consequently, production tends to remain below the social optimum. In an effort to approach the socially optimal quantity of public goods and services, a public authority is usually established by Congress or another legislative body. The purpose of such an authority is to maintain an appropriate quantity of unspoiled (or reclaimed) mountain slopes, clean rivers, and healthy work environments. These authorities attempt to meet their goals through regulation and the enforcement of standards.

That the market fails in allocating resources efficiently whenever externalities and public goods are involved is particularly true in the coal-mining industry, in which production began to expand significantly in the 1970s. A resurgence of public awareness of environmental quality increasingly pressured legislators to come to grips with the issue of inefficient allocation. What followed were the passage of the Federal Coal Mine Health and Safety Act (FCMHSA) in 1969 and the amendments to the 1955 Clean Air Act (CAA) in 1968, 1970, and 1977; the establishment of the Environmental Protection Administration (EPA) in 1970 and the Surface Mining Control and Reclamation Act (SMCRA) in 1977; and approval of environmental legislation by various local and state governments. The consequence of all this new legislation, which was enacted within a relatively short time period, is a maze of regulations and control that is still in an embryonic stage of development.

REDISTRIBUTING EXTERNAL COSTS

Environmental impacts in coal-producing regions are not only an externalized mining effect but also a localized one. Not the consuming but the producing regions are affected most, and unless the effects are accounted for by regulation and enforcement, environmental problems burden those who participate directly or indirectly in the extraction of coal and not those who benefit from its use. Moreover, the burden appears to be regressive when related to income because most coal-consuming regions have higher income levels than the coal-producing regions.

One can view unmitigated environmental impacts, regardless of whether they affect human safety, land quality, or highway usability, as a tax on mining area residents because they are a burden residents cannot escape unless they exclude mining activity from their region. For obvious economic, sociological, and political reasons, this is not a feasible alternative; but until the late 1960s, the residents accepted the burden, materially lowering their real income and well-being.

New safety and environmental legislation at both state and federal levels materially altered this situation. Statutory limits were placed on mining activity in terms of the use of certain capital equipment and on the mining activity's surface effects, such as water extraction and emission, land degradation, and road use. All of these were intended to prevent the emergence of externalities by internalizing them into the production process so that the private costs were more in line with the social costs of mining coal. However, serious administrative, legal, and political issues often stood in the way of realizing this goal.

With respect to establishing limits on the environmental impacts of mining, the issue of how much limitation is crucial. The beneficiaries of reduced environmental degradation are generally not those who create it; hence, they have every incentive to call for zero adverse impacts because this would maximize welfare from their point of view. From the standpoint of society, however, because environmental impacts from mining are always present in one form or another, the objective is somewhat different. The relevant question is not how much environmental damage is acceptable to a specific group, but rather what amount of damage corresponds to the point at which the marginal cost of internalizing or controlling it is just equal to the marginal social benefit obtained.[1] In other words, the level of adverse environmental impacts should be reduced so long as the marginal benefits (cleaner air, reclaimed land, safer working conditions) exceed the marginal costs of achieving them.

REGULATORY STRATEGY AND BEHAVIOR

There exist many different forms of regulation and control. To review them all insofar as they affect the coal industry is beyond the scope of this book. Only the major forms of regulation and their impacts are examined in the following sections.

In order to appreciate fully the complexities of the regulatory process, one must first understand the typical behavioral patterns of regulatory bodies.

If environmental property rights were clearly defined and enforceable, any two parties—for example, a polluter and a victim—could through bargaining resolve differences over the effects of pollution on the value of property. One would expect that differences would be settled by relying on market transactions which could lead to an optimal solution. But in fact, environmental property rights are poorly defined and rarely enforceable, and bargaining is used infrequently. Consequently, the market is unable to allocate environmental resources efficiently. Whenever this situation occurs, regulatory bodies are created to serve as surrogates for the market. The Interstate Commerce Commission, Federal Drug Administration, Mine Safety and Health Administration, Office of Surface Mining, Environmental Protection Agency, and Civil Aeronautics Board, are good examples.

Several strategies are available to regulatory bodies for intervening in the market. First, charges could be levied on users of the environment for discharge of wastes or creation of detrimental effects. Depending on the capability of the environment to assimilate the waste or effect—acid runoff, unreclaimed land, hazardous working conditions—charges could vary in accordance with the marginal damage created by each unit of waste or effect.[2] For several political and administrative reasons, however, this strategy, which is considered by many to be the best strategy, is rarely used.

A second strategy represents a piecemeal approach to environmental control. Public bodies often set quality limits on the use of fuels—a maximum percentage of sulfur in the oil or coal burned, or, for the use of roads, a maximum weight per truck axle. So long as the initial flow of fuel or trucks, in our examples, remains constant, environmental quality can be preserved. If, however, through expanded use more fuel is burned or more trucks use coal-haul roads, environmental quality will decline despite the legislated limits. The piecemeal approach can be and is used selectively, but it is not a comprehensive or optimal solution.

The third and most widely used strategy, although not necessarily the best, is government regulation and containment of effects on the environment. Federal, state, and local authorities issue licenses, permits, and sometimes quality standards that must be observed by those whose activities influence the environment. In some cases, firms are allowed to select their own technological mix in meeting standards, but in more instances than not production, reclamation, and safety-training methods are prescribed by law or regulation. The problem here is that what may be optimum for one firm or mining region may be less so for another, so that minimum production costs are often not attainable. Consequently, final product prices may once again fail to reflect the true social resource costs of producing them.

The enforcement of regulation. To some extent, all three strategies mentioned require enforcement. The ideal would be voluntary compliance, which would be vastly less expensive. It would also obviate the need to use court procedures to adjudicate conflicts, a definite benefit because the legal system is an excruciatingly slow, expensive, overburdened, and uncertain enforcement tool, a process in which legal technicalities often override substance.

Still another reason for relying on voluntary compliance is that regulation and control in conjunction with due process can create "perverse incentive" behavior. A violator can decide that it is less expensive in the long run to fight a licensing requirement or reclamation ruling in the courts than to comply with it. One may conclude, for example, that it is to his advantage to surface mine without a permit and to fight any desist order in the courts rather than to seek a mining permit. By the time a verdict is rendered most of the coal will have been extracted from a site, and even after paying legal expenses and possible fines, the acquired return will be sufficient to have made the venture a financial success. Clearly, voluntary compliance would have been preferable in a case like this.

Finally, it must be recognized that regulations and their enforcement are subject to political whim and influence. Outcomes of controversy between regulator and firms are strongly influenced by relative bargaining strength, including political influence. More often than not, issues are resolved in favor of those being regulated. Also, over time, regulatory zeal and effectiveness diminish. A regulation's long-term effect, therefore, is less certain, particularly when contrasted with the durable nature of effluent or other environmental damage charges.

Politics and decision making. Borrowing from the description of a model of perfect competition, we know that a model of a perfect political system would, by analogy, include perfect knowledge for policymakers and voters. In the competitive model, access to the market is guaranteed, and sellers and buyers are held accountable for their actions through gains or losses in profits and utility. The ideal political model likewise guarantees access to the decision-making process through voter access to the elected official: the official is accountable to the constituents, who may place their votes elsewhere in the next election.

Unfortunately, neither the marketplace nor the political system is perfect, and elected representatives, the decision makers, find ample opportunity to escape accountability and reduce the public's accessibility to them. They do this in a number of ways:

- by shifting down the line the responsibility for making controversial decisions, for example, from federal to state, or from state to local levels, or from Congress to the federal or state bureaucracy
- by looking for alternative policies whose costs are hidden or fall on the least influential groups of the electorate
- by postponing decisions

In all of these cases, political accountability and public access to decision makers is reduced, and the status quo endures. While there is virtue in moderation and gradualism, there is also the risk that undesirable environmental practices will become institutionalized, that their unmitigated effects will grow exponentially, and that social welfare will suffer irreversible setbacks.[3] A basic justification for the existence of the political system, the need to reconcile and resolve conflicting public interests, thus can be seriously compromised.

The role of regulatory bureaus. If it is true that elected officials frequently consign the task of controversial decision making to others, it becomes important to examine the behavior of those on whom the ultimate decision-making burden falls. In the discharge of their legislative duties, elected representatives often set broad policy goals only, leaving it to their respective bureaucracies to develop specific standards and procedures.

A bureaucracy faces a unique buyer of its services and many sellers of the resource inputs it needs.

1. The buyer of regulatory services. Unfortunately, the bureau's customer is often not the public but the sponsoring authority that supplies the bureau exclusively with annual funds.[4] Thus the demand function facing the bureau originates with its sponsor—the legislature—and is not a direct reflection of the marginal social value of the service rendered to the public. One can view this relationship as a bilateral monopoly: there is one demander, the sponsoring agency, and one supplier, the bureau. The social utility function, the people's view, enters into this relationship indirectly and only insofar as legislators correctly approximate and convert this view into legislative policy criteria.

Because bureaus do not sell services in the marketplace, the traditional concept of price per unit of service has no meaning. Exchange takes place only between bureau and sponsor; for the service rendered, the bureau receives its budget. Not unexpectedly, to maximize this budget often becomes the objective of the bureau. Poor service is punished by budgetary cuts, and good service is rewarded with budgetary expansion.

2. The sellers of factor inputs. Very much like other organizations, the bureau buys its inputs at competitive prices in the market. Thus it competes with other buyers for factor inputs and, other things equal, must pay the governing market price for labor, services, and materials. In the factor market the bureau behaves no differently from other entities.

As a rule, bureaus make the necessary expenditures to provide service during a single budgetary year. The segmentation of long-run funding into annual increments is somewhat more pronounced in a bureaucracy than in a private enterprise. This often introduces a rather short-term perspective to regulatory activity and to the measure of its success. It would be better to judge the importance and success of regulation not in terms of the short run, but on the basis of its long-term impact on society and the physical environment.

Most observers of bureaucratic behavior would agree that the majority of bureau workers are sincerely interested in serving the public well and that they are motivated to do what is best for the citizenry, for the industry involved, and, by way of doing what is best for the citizenry, for the sponsoring authority. But, like anyone else, bureaucrats have personal motivations. One of these is the growth of responsibilities. This can be achieved in two forms—intensified or expanded service by the bureau. In either case, the objective is the temporal maximization of budgetary appropriations. And because the source of expanded funding is the public authority, relations between it and the bureau sometimes become awkward, characterized by threats and retreats, the playing of bureaucratic games, and appeals to common objectives.

One way for a bureau to obtain higher funding is to specialize in services that yield economies of scale. By doing this, it can use unspent (saved) funds to branch out into other services. A bureau also may seek to specialize in services for which there are economies of joint production. This, too, would release funds for other activities. Because bureaus regularly face general uncertainty, they often attempt to broaden their service line to hedge against it. This also protects them against unexpected factor cost escalations and against uncertainty in the demand for their services.

Bureaus are responsible to one sponsor only, but often they attempt to expand their range of services. This can lead to jurisdictional conflicts between bureaus and to unnecessary duplication of regulatory service at one level of government or between levels. In either case, once discovered, such duplication should be removed, despite the existence of vested interests.

REGULATING BY PERMIT

One of the principal regulatory tools in the coal industry is the permit to engage in mining. A permit is defined as an authority to operate issued by a public agency. Recurrent reviews or inspections are excluded from this definition because they are more properly viewed as enforcement tools designed to safeguard the integrity of previously issued permits.

Permits are granted by regulators who use them to enforce explicit standards for entry into and continuation in an industry. Producers who fail to meet the minimum standards are barred from entering; they are undercapitalized or inexperienced or have previously demonstrated that they do not adequately safeguard environmental and human resources.

Over the past fifteen years, new safety and environmental legislation has raised considerably the cost of entry into the coal-mining industry, but not to the exclusion of those who are able and willing to make the necessary expenditures to meet minimum standards. The result has been a reduction in the number of wildcat (marginal) operators who find the risk of operating without permits or the cost of legal entry too high.[5]

Because all applicants who satisfy minimum standards are granted permits, the number of new entrants depends, first of all, on the number of applicants. The number of applicants, in turn, depends on the expected rate of return on investment in new mines, and the number of departures from the industry depends on the profitability of existing operations. The link between entry and exit in the coal industry, however, must be regarded as stochastic because regulators who issue permits understand incompletely how a potential investor's expectations are related to the number and kinds of permits required and to actual profits.

As many as 100 permits may be required at one time or another to plan, construct, operate, and close a coal mine. The exact number will vary with differences in county and state regulations and with the characteristics of the applicant. More than half of the permits are issued by state bureaus; local bureaus issue the smallest number.

A considerable amount of jurisdictional fragmentation pervades the regulatory process. This appears to be more a result of bureaucratic convenience than the consequence of a desire for efficiency in permit administration and service to applicants. This is not particularly surprising because bureaus depend for funding on different sponsors, some located at the federal level and others at the state and local levels.

The economic consequence of jurisdictional fragmentation in the issuance of permits is a higher acquisition cost for permits. The higher costs are derived from both the side of the bureau and that of the applicant. Bureau location and distance from the applicant and piecemeal filing procedures, first to the federal and then to the state level or vice versa, result in higher engineering and bureau review costs, and possibly in a duplication of effort. Combined, they raise the aggregate costs of obtaining permits.

There appears to be little to commend the existing fragmented procedure for obtaining permits. In the United States the development of regulatory systems appears to be based on the tradition of fragmentation. From the standpoint of those being regulated, however, and in terms of efficiency, the structure of the system could be improved significantly.

One way of improving regulation in the coal industry is through simplifying the process along functional input lines. Not unlike other production processes, coal mining incorporates the use of three basic inputs: human, land, and capital resources. Restructuring the permit and enforcement system to address specifically the input functions of these three resources could greatly simplify the preservation of

human and environmental integrity. For instance, permit requirements dealing directly with human resources could be grouped under
the aegis of one public authority responsible for the issuance of all
relevant permits. Likewise, all permits that influence the preservation of natural resources—land, water, vegetation, and air—could be
integrated under the umbrella of a single authority. A similar procedure could be developed for maintaining the safety standards of the
capital equipment used in mining.

The advantage of structuring the permit system along functional
input lines is that the effectiveness of the entire system can be
monitored and evaluated. That is, the contribution that individual
permits make to the preservation of social welfare or to the
maintenance of an unspoiled environment can be clearly identified
and analyzed. In the absence of such an integrated functional approach
to permit issuance, regulatory responsibility is widely scattered
among many jurisdictions and agencies. The system in use now
makes it extremely difficult, if not impossible, to estimate accurately the regulatory and compliance cost and effectiveness of providing
safety in deep mining, surface-mined land reclamation, or unpolluted
streams.

SIMPLIFYING THE PERMIT-ISSUING PROCESS

Restructuring the permit-granting system along functional lines and
integrating local, state, and federal permits by function appears to be
a useful, but by no means simple, first step. Following functional lines,
three bureaus could be created to issue permits for the utilization of
human, land, and capital resources in coal mining. Each bureau would
be changed with all local, state, and federal responsibility for permit
approval. For example, a new bureau for regulating human resource
use in coal mining would issue permits to employer applicants.
Similarly, a bureau to handle land use permits and another to evaluate
capital equipment permit requests would be created.

The suggested simplification of the permit process understand-
ably would elicit opposition from existing bureaus and their sponsors. The new format would require the relinquishing of jurisdictional
and functional responsibilities, something bureaus are usually loath
to do. But the result would improve program operations and reduce
operating costs. Under the new structure, an applicant would have
a clear notion of what permits are required for entering or continuing in coal mining. Because fewer jurisdictional entities would be involved, fewer studies on engineering, land use, the environment, and

safety, would be needed to support an application. Also, presumably these studies would be prepared just once at the outset, thereby saving time and avoiding possible duplication. Inevitably, compliance costs would decline from present levels.

Bureau review costs also would be lower under an integrated functional approach because only one master bureau would be needed to review proposals for labor, land, and equipment use in mining. Review time would be shortened because each jurisdiction would have access to every other jurisdiction's evaluation, thus reducing the time interval between filing and approval or rejection. Professional staffs could be used interchangeably among bureaus.

Finally, an integrated functional approach to issuing permits would make it possible to evaluate the cost and success of each phase of permitting and its contribution to the overall objective of safeguarding resource use. If the costs and effectiveness of each element in the permit process were identifiable, the success of the entire program could be judged more accurately.

It is axiomatic, yet often overlooked, that governmental expenditures of all kinds, including those on the permit system, should be judged on the basis of their impact on people and the environment. Arriving at a sound judgment is made extremely difficult, however, by the fragmentation of the permit process now in use. A program approach organized along functional lines would alter this and would provide the necessary information on the causal linkages between program inputs and program impacts.

There is little doubt that the difficulties associated with restructuring the permit system are real, but they should not be used as an excuse for avoiding rigorous analysis of the issues. One problem of analysis that will surface is the fact that government programs do not entail market prices. Therefore, it becomes necessary to impute the prices or values of programs. The practice is known as shadow pricing and has a tradition of at least two decades in economic analysis.[6]

With the aid of shadow prices it is possible to use cost-benefit techniques for analyzing the permit system across subject areas.[7] For example, when the ratio of benefits to costs of a program, such as all safety training in mining, is 1.0, costs are equal to benefits. As the ratio increases, the benefit per unit of expenditure accruing has increased. In theory, if a land reclamation program yielded a ratio of 1.7 and a safety training program a ratio of 2.5, then using the criterion of economic efficiency to adjust for program size, the government could be advised to favor safety training over land reclamation.

Because it is intended to have an impact on the structure of the permit-issuing process, program analysis may be abused in the decision-making process. A study of an existing program may reveal that it is largely unsuccessful in producing expected outcomes. This may stimulate a request for enlarged funding and effort. In order to prevent participants in the regulatory process from finding bases for expanding their program regardless of study results, it is important to include an analysis, and wherever possible, an estimate of the marginal results of additional expenditures.

Having a functional program structure in the permit process would reduce the complexity of the existing system. In a systematic classification of all governmental permit-issuing activities, the causal linkages would be displayed between expenditures and benefits. Multiple-year plans could be developed within this framework to bring together program and resource data and to illustrate regulatory activity costs and benefits. To a baseline year can be added over time incremental program changes to expand or contract the regulatory domain. From the standpoint that it attempts to identify all permit programs—federal, state, and local—and the links among them, program analysis, and ultimately program budgeting, are comprehensive and rational.[8] Whatever choices among programs cannot be readily transformed into dollar terms can be made on the basis of political and social values.

The overriding objective of requiring permits for the use of labor in mining is safety. This also seems to be the dominant objective of permit regulations for the use of equipment. Government activity attempts to provide greater safety by regulating human behavior through training and proper equipment use and also by regulating the nature of the equipment used in mining. Both regulatory functions are based on a common denominator: greater safety for miners. It is, therefore, conceivable that at some time in the future, permit issuance activities for human and for capital use in mining would be combined into one bureau function whose mission would be to maximize human safety through efficient use of the permit process. The only function remaining to be established would be a permit program designed to preserve the integrity of land use. Ultimately, therefore, one can envision but two regulatory programs: one that concerns human safety, broadly defined, and one that governs land use and environmental preservation.

So far, the discussion has centered on the three major input functions in coal mining. There exists, however, a fourth important activity—the transport of coal from mine to tipple and tipple to consumer, and occasionally from mine to consumer—which is the subject of considerable regulation. Once again, these regulatory activities attempt to safeguard people and the environment. And although these regulations could be classified under one or the other function, it seems more appropriate to keep separate the regulatory activity of coal transport. In a strict sense, transportation is not part of the production process. It becomes important only after the coal has reached the mine mouth. A strong case can be made, therefore, for considering activities related to coal transport separately from those that have a direct impact on coal production.

EVALUATING REGULATION OF COAL MINING

Over the past twenty years, Congress has mandated that its rules to maximize public welfare be implemented through regulatory decision making. New bureaus such as the Environmental Protection Agency, the Office of Surface Mining, and the Mine Safety and Health Administration have considerable freedom in translating the congressional mandate into action.[9] Not unexpectedly, ambiguity, indecision, and inconsistency have developed in the coal-mining regulatory network. As with most public agencies, one reason for this is the absence of an objective market test for valuing public services. Another cause is the embryonic stage through which these new bureaucracies and their regulatory structures are just now passing. In time, as these bureaucracies mature and their decision-making processes become institutionalized, it will become more evident what sort of criteria and weights they use in reaching decisions. In the meantime, as the discussion of the previous sections suggests, the permit issuance process ought to be streamlined and simplified along functional lines to improve service and make it easier to evaluate program effectiveness.

One way to evaluate the usefulness of the suggested change is to compare the conditions that probably would result from the change (a positive approach) with the optimal condition that would be developed from normative theory. This is a difficult task, but not insurmountable, and it should be a serious contender among subjects for future research.[10] Regardless of the findings of such research, even if the new permit structure does not generate conditions identical with those considered optimal, this alone is not a sufficient argument for rejecting the proposal; the results may still be better than those that flow from existing procedures or from feasible alternatives.

It is plausible that the existence of potential conflicts of interest between legislators and regulators, and among regulators themselves, will retard the process of achieving a permit system organized along functional lines. But if regulatory policy is to serve the will of the public and of the legislature more effectively, such conflicts of interest must be resolved. This can be accomplished best through the creation of joint public-business-labor-governmental (legislative and regulatory) boards to arbitrate conflicts and sponsor solutions.[11] There is nothing moral, immoral, or amoral about any particular degree of public-business-labor-intragovernment cooperation. The adversary relationships that characterize much of the interchange among these groups today seem artificial and unnatural. They rarely surface during periods of war or other national crises. There exists no a priori reason for them to surface during normal times.

VI

ENVIRONMENTAL AND HUMAN ISSUES

COAL UTILIZATION AND AIR QUALITY

The burning of coal, like that of other fossil fuels, leads to the emission into the atmosphere of residuals that have a detrimental impact on the environment. The emitted elements include carbon dioxide (CO_2), water vapor, sulfur dioxide (SO_2), nitrogen oxides (NO_x), carbon monoxide, and hydrocarbons.

The most troublesome element is SO_2. Its emission, first of all, affects human health. It is believed by some, although the scientific evidence is far from conclusive, that when SO_2 concentrations are much higher than normal ambient quantities, physiological responses occur in the long run. It is thought that there also exists some interaction between SO_2 and other pollutants, including particulates and SO_2 derivatives, which makes long-term exposure to SO_2 hazardous. Sulfur dioxide emissions also affect plant life, including crops. It is very difficult to estimate the actual costs of crop damage. One source puts the value of crop damage in the eleven Western European members of the Organization for Economic Cooperation and Development (OECD) at approximately $500 million per year in 1981 prices.[1] In the United States, several electric utilities burning high-sulfur coal have agreed recently to compensate farmers for crop losses sustained from uncontrolled sulfur emissions. Finally, SO_2 emissions are held responsible for corrosive damage to metal and man-made concrete, marble, and limestone structures.

In addition to SO_2, emitted nitrogen oxides (NO_x) formed during the combustion of coal affect human health and plants and add to acid precipitation. The extent of these effects is largely unmeasurable and certainly not as great as those from SO_2. It is, however, clear that the increased combustion of fossil fuels has led to larger quantities of SO_2 and NO_x being emitted, accompanied by significantly higher concentrations of acid depositions. For many parts of the world, especially the east central European countries, Scandinavia, the northeastern United States, and eastern Canada, this has become an issue of considerable concern.

In the United States, concern over increasingly serious air pollution found legislative expression in the Clean Air Act of 1963. It took another seven years before the federal government empowered the Environmental Protection Agency in 1970 to establish uniform air quality standards. These ambient air standards set maximum emission levels for the seven most common pollutants: sulfur dioxide, nitrogen oxides, particulate matter, lead, carbon dioxide, hydrocarbons, and ozone. Concerning coal-fired furnaces, the act limits the emission of SO_2, NO_x, and particulates.

In 1971, the EPA issued source performance standards for new coal-fired utility boilers larger than 73 megawatts.[2] These standards prohibited utilities from emitting more than 1.2 pounds of SO_2 per million Btu (MMBtu) of heat input. The country was also divided into 274 air quality control regions. Where the ambient pollutant concentrations were below or equal to air quality standards, the region was labeled an attainment area. Where they exceeded the standards, a nonattainment area designation was assigned. Utilities were required to monitor continuously SO_2 emission levels in the flue gas outlets of coal-fired boilers.

Some of these emissions are natural. They arise from sea evaporation, volcanic eruptions, and organic decomposition. In the eastern United States, however, only about 10% of the sulfur emissions are believed to originate from these sources. The rest are attributable to electric utility combustion, industrial fuel combustion, industrial production processes, and transportation. The source distribution of the SO_2 emissions is shown in table 63. At least 62% of the sulfur emitted into the atmosphere has its source in coal combustion. Substantial progress can be made in reducing overall sulfur emissions by controlling the way coal is burned and by using cleanup methods (called scrubbing) on flue gases.

In 1979, revised new source performance standards were issued by the EPA. These built on the 1971 standards by requiring that all boilers purchased new or modified after 1978 use coal with reduced sulfur content. Utilities were required to have removed at least 90%

of the SO_2 from the emissions of coal-fired furnaces unless these emissions contain less than 0.6 pounds of SO_2 per MMBtu of coal burned; in such cases scrubbing was declared unnecessary. Irrespective of emission levels, however, sulfur removal was required to be 70%-90% effective. The revised standards also require the use of flue gas desulfurization (FGD) techniques for all new or modified utility boilers because it is contended that only these techniques can reduce emissions by more than 70%. The intent of the agency standards was to encourage utilities to use the best available control technology (BACT) to reduce emissions not only of SO_2 but also of NO_x and particulates. This policy generated considerable controversy.

The strongest endorsement for BACT came from midwestern and Appalachian supporters who feared a strengthening of the shift to western U.S. coal production, and with it a steep decrease in Appalachian mining activity. An additional concern, of course, was air pollution. It was argued that BACT would ensure that newly built utility boilers would reduce air pollution while the old standards might not have. This argument, however, is of questionable merit because the only way to guarantee that fewer total pollutants are emitted into the atmosphere is to control their quantity, not the way they are separated from coal.

It was also hoped that BACT would reduce land degradation in the West by decreasing surface mining there. If all coal, regardless of sulfur content, had to be scrubbed, it made little sense to buy low-sulfur, high-priced western coal in place of the high-sulfur, lower-priced midwestern coal. On this point western coal interests lobbied successfully for a sliding-scale approach to emissions reduction that would give them some market protection. Scrubbing their coal to a 70% effectiveness would still be less costly than scrubbing high-sulfur coal of 90% of its harmful emissions.

Ideally, it would seem best to set standards for ambient air quality and leave it to the individual utilities to select the most appropriate technology with which to clean their coal and meet the standards. How these standards are attained is unimportant so long as air samples taken from areas surrounding power plants meet them. Utilities could attain these standards by burning only low-sulfur compliance coal; they could attain them by mixing compliance and non-compliance coal to a given emission level; or they could install electrostatic precipitators or bag houses.[3] In the past, precipitators have not proved as effective in capturing particulate matter from low-sulfur coals as from high-sulfur ones. Consequently, attention has shifted to bag houses. To date, these have shown a capability to capture routinely 99.9% of all particulate matter in the flue gas—a proportion well within existing standards. One virtue of the bag house method is that it can capture both particulates and SO_2. In only ten years this technology has proved to be a viable and economically attractive pollution-control option for utilities.[4]

Most U.S. electric utilities located in the Midwest and the East burn high-sulfur eastern coal to produce electricity. In recent years their interest in bag houses has risen sharply, particularly when bag houses are used in connection with an all-dry sorbent injection process. In this process a sodium-based reagent is pulverized and fed into the flue gas stream prior to the stream's entering the bag house. The agent collects on the bags, and as it attempts to pass through it removes the SO_2. This process is particularly promising for use with more stringent SO_2 standards that might be introduced in the future.

There is considerable excess capacity in the utility industry at this time. Utility companies have little incentive to expand capacity or even to convert existing plants to accommodate the use of coal. The North American Electric Reliability Council estimated that over one-half of the new coal and nuclear power plant capacity scheduled for 1979 through 1988 had been delayed and that two-thirds of the operating plants that could be converted economically from burning oil to coal still use the former.[5] Utilities apparently are not currently overendowed with cash or the prospects of a rosy future.

Several factors are responsible for the reluctance of electric utilities to expand or convert existing capacity. Energy conservation at the residential, commercial, and industrial levels has been unexpectedly effective, reducing growth in the demand for electricity significantly over the past ten years. Recent economic recessions have further diminished this demand. More frequently now than in recent years talk is heard of time-of-day pricing of electricity, the use of which can translate into major shifts in load curves for utilities and can reduce the need for additional future capacity even more. Finally, the high cost of capital has greatly reduced utility executives' willingness to enter into expensive capital construction and modification projects. Coal conversion is being avoided and new coal-fired furnaces are not being constructed at this time because electric utilities do not feel they can afford them without government subsidy. The long-run implication is that future electric energy prices will be higher than they would have been if capacity had been expanded and furnaces converted. This discourages the introduction and use of energy-intensive equipment and may downwardly bias prospective gains in productivity.

Another factor discouraging expansion is U.S. Senate Bill 3041. Introduced in late 1982 by a group of senators to amend the Clean Air Act, the bill proposed the establishment of an acid deposition impact region consisting of thirty-one states, located mostly east of the Mississippi River, and the District of Columbia. The bill provided for a general reduction by 1995 of SO_2 emissions in the region by 8 million tons per annum from 1980 levels.[6] If an increase in SO_2 emissions from a stationary source were observed after 1981, it would be required to be offset by an equal reduction from another source by 1995. Stationary sources of SO_2 emissions put into operation after 1995 would be accompanied by reductions in emissions elsewhere in the region, unless the new facility used the best available control technology.

In their efforts to meet existing or new air pollution standards, electric utilities first of all can avail themselves of standard flue gas desulfurization (FGD) techniques. Additionally, they can decommission or reduce the use of plants burning high-sulfur coal, buy mostly low-sulfur coal, equip their boilers with new FGD systems, or buy, where feasible, electric power from other utilities. Whether electric utilities can be retrofitted with new FGD systems depends on many factors, the most significant of which are age of equipment, physical constraints, and unit size.

So far during the 1980s practically no retrofitting of boilers has taken place. This fact is directly attributable to the lack of growth in the demand for electricity and to uncertainty regarding the future costs of transporting coal from mine to boiler. During the years immediately following the oil boycott, the demand for electricity continued to grow vigorously. As time passed and manufacturers and consumers adjusted their consumption under the higher electricity rates, however, average annual growth began to moderate and actually became negative. In 1982, for example, power generation fell by 2.3%.

The utilities' alternative to retrofitting, burning low-sulfur coal, is of course of great interest to coal producers. The industry would be greatly affected if SO_2 emission standards were implemented uniformly and actually enforced. Professor Zimmerman of the MIT Energy Laboratory has constructed a dynamic forecasting model that includes estimates of these industry effects.[7] We will examine his estimates of future distributions of U.S. coal supplies under the assumption of full compliance with a maximum allowed emission level of 1.2 pounds of SO_2 per MMBtu beginning in 1985. He assumed that factor-input costs would remain constant.

The model predicts an acceleration of western coal production, as shown in table 64. By the year 2000, low-sulfur output from Montana and Wyoming will have increased to more than five times its 1980 level. Appalachian coal can be divided into two coal supply regions: the northern, which produces mostly high-sulfur coal, and the southern, which produces a low-sulfur output. Northern Appalachian production, after a temporary decline and subsequent surge, will settle back roughly to its 1980 level; southern Appalachian production will decline significantly from 1980 levels after a brief surge. The short-lived decline in output from the Midwest is dramatic, as is its recovery and growth in future years.

The explanation for these changes is that the stringent air pollution standards will initially make clean-burning southern Appalachian coal more attractive east of the Mississippi River. Northern Appalachian and midwestern coal will become less attractive. In the midwestern demand regions, southern Appalachian and western coal will be substituted for coal from the Midwest and from northern Appalachia. But as southern Appalachian reserves become depleted, the price of this coal will increase, a fact which sets the stage for a reverse substitution of regional coals and the installation of sulfur-removing equipment at consumption points. Northern Appalachian and midwestern producers, holding large coal reserves, will recapture some of the markets they previously lost to low-sulfur coal producers.

The markets currently supplied by midwestern producers seem to be the most sensitive to SO_2 legislation. The eastern markets are more insulated: little, if any, western coal is sold in these markets, and little is expected to be under new enforcement of standards. The new SO_2 legislation has the effect of making southern Appalachian coal more attractive to all markets in the short run. As depletion pushes up mining costs, however, northern Appalachian and even some midwestern coal will become a more attractive alternative, even after the utilities allow for the additional costs of purging the coal of its sulfur. Over time, the Zimmerman model shows, eastern consumers will continue to prefer eastern coal (that produced in northern Appalachia), and the largest share of southern Appalachian output will be dedicated to metallurgical use and to the export market. Ultimately, the Midwest will be the most sensitive to pollution standards, with the West affected very little and the East not at all. In Appalachia, only intraregional redistributions of production will occur.

Actual output statistics displayed in table 13 show that while the surge in Montana and Wyoming coal production is already dramatic, the expected decline by 1985 in midwestern production has been relatively modest. The 1.2 SO_2 standards are not uniformly in use yet, however, and emissions inspections and control are not being pursued vigorously. A considerable amount of noncompliance coal continues to be burned without regard to prescribed SO_2 emission standards.

It probably will be a decade or longer before the results of the clean air legislation already passed or now being considered will have a measurable impact on air quality. Most of the SO_2 pollution today originates from older plants built prior to the enactment of clean air legislation. Even though many plants are aging and have outlived their intended lifespans, they are being repaired to avoid the expense of building new ones. From 1974 through 1981 alone, the construction of 39 new coal-fired units was canceled and plans for 335 were delayed.[8] Uncertain future demand for electricity, high interest rates, and losses from ventures into nuclear power plant construction make it extremely difficult, if not impossible, for utilities to build clean, new coal-fired facilities. The Environmental Protection Agency predicts that SO_2 emissions will continue at present levels into the foreseeable future.[9]

THE GREENHOUSE EFFECT

In primordial times the Earth's atmosphere contained no free oxygen. The atmosphere was transparent to ultraviolet radiation; life then existed only in the sea. The sea's microorganisms began to develop the photosynthetic process of forming carbohydrates and at the same time released oxygen into the atmosphere. These oxygen molecules slowly accumulated and then decomposed into free oxygen atoms, which combined to form molecules of ozone (O_3). Once established in sufficient quantities in the higher layers of the atmosphere, the ozone molecules, capable of absorbing ultraviolet light, began to act as a stratospheric shield to filter out harmful rays. This shield allowed the formation of photosynthetic plant life on Earth and led to the development of today's atmosphere.

The atmosphere protects the Earth from radiation harmful to contemporary life forms, and it also acts as an insulator. The Earth's surface radiates infrared light, which is absorbed by molecules of carbon dioxide and water in the atmosphere. The absorbed radiation is re-radiated in all directions, and some of it falls back on the Earth's surface and keeps it warm. All fossil-fueled power stations and all engines that use carbon-based fuels emit carbon dioxide as well as water vapor. About 50% of the carbon dioxide is retained in the atmosphere. The greater the amount of energy produced and consumed, the greater the emissions of carbon dioxide and water vapors into the atmosphere and the greater the amount of infrared light and heat retained. With increased emissions, there is the danger that the Earth's surface will warm to a point at which the balance of all of its life systems will be endangered. The threat has been called the *greenhouse effect*.[10]

The main source of the increase in atmospheric carbon dioxide in modern times has been the combustion of fossil fuels. Concentration levels have risen from 280 parts of CO_2 per million parts of air (ppm) to 340 ppm in this century. If the level reaches 660 ppm, perhaps by the end of the next century, the Earth's surface temperature will have increased approximately by 3° C. In the Northern Hemisphere, the rise could even be 4° to 6° C, which would induce polar ice melts and coastal flooding. Also altered would be rainfall patterns, the sensitivity of land masses to desertification, fish stocks, forestation, and water supplies. The restructuring of social, economic, and political environments would surely follow. In a study supported by the National Science Foundation, several scientists examined the impact of a range of energy supply and demand scenarios on CO_2 levels.[11] Their conclusion was that if CO_2 levels are not to exceed 420 ppm by the year 2050, total energy consumption and the use of fossil fuels, especially coal, ought to be curtailed. Rather than suggesting hasty restrictions on fossil fuel and coal consumption, however, the scientists recommended improved efficiency in the use of energy that could postpone the global CO_2 problem for centuries. Greater efficiency means technological improvements in energy-using machinery and equipment, technological advancements in the use of solar photovoltaic systems and nuclear fusion, and biomass exploration. They predict that the world will increase its dependence on electric power because such power delivers clean, convenient, and highly controllable energy to the final consumer. Electricity is particularly well suited for use in our "high-tech" age and is compatible with all energy sources of the future.

SURFACE MINING AND LAND RECLAMATION

When land is surface mined without adequate reclamation, environmental damage usually develops. Such damage can affect many aspects of life: economic, social, and aesthetic. It takes the form of erosion of topsoil, sedimentation, water pollution and increased runoff, landslides, air pollution from dust and debris, loss of natural beauty, and disturbance of wildlife and vegetation. Economists label such potential damage created by surface operators during the course of mining as *externalities.*

The oil crisis of the middle 1970s focused considerable attention on coal as a fuel substitute. Because high percentages of all U.S. coal output are mined on the surface, a conflict has arisen between the need to preserve environmental integrity and the need to maintain a continuous flow of energy. Nowhere have the external effects of coal mining been more apparent, or more directly felt, than in the coal regions of Appalachia.[12] As practiced in the hilly terrain of the region, surface mining has caused all of the problems listed above.

Surface operations around the world differ greatly from country to country. English surface operations, which are small-scale and short-lived because coal seams lie close to the surface, are vastly different from those of Germany, which are conducted on a massive scale. In the United States a similar contrast is found between Appalachian surface operations and their western counterparts. In a recently published study, the OECD found that in Poland less than 0.1 square kilometer of land is disturbed for every 1 million tons of coal surface mined.[13] By contrast, in other countries up to 8 square kilometers of land are needed to surface mine an equal amount of coal.

The two principal techniques of surface mining are area and contour mining. A recently practiced and less widely used method is mountaintop removal. The first technique is used when land is relatively flat or gently rolling, as, for instance, it is in West Germany's Ruhr area; here 175 square kilometers of land have been disturbed and 23,700 residents have been resettled during the mining of brown coal. In the hilly southern Appalachian region, on the other hand, a dense hardwood forest cover makes area mining infeasible. Contour mining, and lately mountaintop removal, are the two most widespread techniques in use there. If reclamation is neglected, land rehabilitation and reforestation may take many decades or, in some instances, may even defy such efforts. A large portion of the Appalachian land surface mined prior to the 1970s remains unreclaimed. Nutrient deficiencies, physical factors, and high concentrations of toxic ions make the reclamation of such orphan lands exceedingly difficult.[14]

In recognition of the adverse consequences of large-scale land degradation, the U.S. Congress in the 1970s began to consider remedial legislation. The first versions of surface-mining control legislation passed by Congress were vetoed in 1974 and 1975 by President Ford, but in 1977 President Carter signed into law the Surface

Mining Control and Reclamation Act (SMCRA). Although passed by the U.S. Senate, subsequent efforts in 1979 and 1980 to weaken the act were opposed by the president and the House of Representatives. In subsequent court challenges, the U.S. Supreme Court interpreted the principle provisions of the act strictly. It seems clear that future changes in the SMCRA would have to be legislative in origin. The only other flexibility exists in statewide differences in interpretation and enforcement of the law. These differences are relatively small and must meet overall federal standards set by the Bureau of Surface Mining, Department of Interior.

The provisions of the SMCRA affect surface- and deep-mining operations. The latter are affected by the act only insofar as their mining activity disturbs surface lands.

Section 515b of the SMCRA identifies twenty-five reclamation requirements that must be met by surface operators. Of these, the most important and far-reaching stipulate that the mined land must be returned to its original use capability and to its approximate original contour. Through revegetation and stabilization, soil erosion and hydrological damage must be avoided and a proper balance maintained in the streams. Topsoil and overburden removal must be managed so as to prevent sedimentation, acid drainage, and soil slides. In an effort to begin the reclamation of abandoned mine sites, the so-called orphan lands, the SMCRA established a reclamation fund supported by a thirty-five-cent-per-ton tax on surface-mined coal and a fifteen-cent-per-ton tax on deep-mined coal. Other provisions of the act prohibit the strip mining of public lands in parks, wildlife refuges, wilderness areas, and national forests. The act also regulates the surface mining of prime farmland and consent rights of surface and sub-surface mineral owners.

Because most surface mining on steep slopes occurs in Appalachia, this region is much more seriously affected by the SMCRA than are the midwestern and western coal-producing regions. Reclamation costs in the latter areas are much lower than in Appalachia. Because reclamation of mined land is mandatory, the additional reclamation costs necessary for restoring Appalachian land can be viewed as a tax on Appalachian surface coal production. Who bears these additional costs, how they are distributed, and who benefits directly and indirectly from reclamation activity are issues that have not been resolved, but some relative estimates have been made.

Within Appalachia, the most strongly affected regions are Eastern Kentucky, Virginia, and West Virginia. These three areas have a much greater share of their minable reserves located under steep slopes than

other states in the Appalachian basin. Table 65 gives an approxima-
tion of the strippable reserves lying under each angle of mountain
slope. The quite tenuous data show a large percentage of Eastern Kentucky and Virginia reserves under steep terrain. West Virginia shows
the third-largest concentration. It is likely that, other things equal,
these three areas will experience a relatively greater impact from enforcement of the SMCRA than will the other Appalachian states. The
impact can be expected to manifest itself in a shrinking share of
surface-mined coal in total production. Evidence bears out this expectation. As table 49 shows, in 1979, the year in which enforcement
of the new standards began, the share of surface-mined coal declined
by 16% in Appalachia. This decline was most strongly influenced by
statistics for Virginia, which showed a decrease of 21.2%; West
Virginia, also with a 21.2% decrease; and Eastern Kentucky, which
had a 15.5% reduction, all states with the highest percentages of their
reserves located under steep slopes.[15] For Appalachia as a whole, surface mining in 1982 remained below the share levels it had attained
in 1978, surely a reflection of the measures taken under the SMCRA.

The primary control mechanism in place to regulate surface mining is the permit system administered by state agencies under primacy
license from the U.S. Bureau of Mines, Office of Surface Mining. To
surface mine, an operator must obtain a permit. In addition to a permit fee, he or she must post a reclamation performance bond to ensure land restoration. The permit application must be accompanied
by engineering plans detailing methods of operation, backfilling and
grading, road access, bench width, highwall slope, revegetation, and
reforestation. While the SMCRA will unquestionably have a significant salutary impact on environmental preservation, there exists
strong evidence that some of the aesthetic consequences of mountain strip mining cannot be reversed.[16]

The federal regulations of the SMCRA create substantial costs
for surface-mine operators, but they also create considerable benefits
for the consuming public. The costs of forcing operators to bear the
externalities created by surface mining are not especially difficult to
measure. Estimating the cost of a damaged environment and the flow
of benefits that arise from land reclamation is much more difficult.

In a study prepared for the U.S. Department of Energy, Professor
Kalt of Harvard University attempted to estimate the costs to surface operators of complying with the SMCRA and the costs to the
public of the externalities created in the absence of the SMCRA. Table
66 shows a summary of these estimates for Appalachia, the Midwest,
and the West. The variation among regions is revealing. Appalachian
costs per ton of coal to comply with the SMCRA and the state regula-

tions are twice as high as Midwest costs and eleven times as much as costs in the West. Compliance costs in Appalachia appear to be more than one-third greater than the externality costs. Because costs exceed the benefits produced in Appalachia, the SMCRA may not be cost effective; that is, overregulation may be the consequence of the passage of the act. For the Midwest, however, the costs of compliance seem to be about equal to the costs of the externalities created, and the cost-benefit ratio is close to one; the provisions of the act correspond closely to the nature of the damage created there. In the West, benefits seem to exceed costs. One critical variable in the Kalt analysis is the reduced aesthetic value to nonlocal people that results from unregulated surface mining. The estimates shown in table 66 ($1.26, $2.00, and $0.93 in the three regions) are quite large and influence the results strongly for the Midwest and West (they comprise 66% and 90% of total externality costs, respectively). For Appalachia, they represent the second largest cost category and only 25% of total cost. If the reduced aesthetic value to outsiders in Appalachia is underestimated, benefits would rise in relation to costs and the SMCRA would be more cost effective. Unfortunately, estimating the pristine value of regions to outsiders is a very hazardous undertaking that requires, in Kalt's own words, "great leaps of faith."

The differences in the cost-effectiveness of the SMCRA for the three coal-producing regions are due to the act's uniform provisions being applied to nonuniform, distinctly different surface-mining regions. The ideal method of preventing land degradation during surface mining would be to implement an optimal tax rate on degradation which would make it profitable for operators to reclaim the land. Unfortunately, information required for implementing such a tax is unavailable. The SMCRA attempts to attain the same objective, not through taxation, but through legislatively established performance standards for the entire United States. There is, however, another way: through the use of an acceptable-standards approach which promotes economic satisficing (in essence, making the best of a bad situation) rather than optimizing.[17] For each mining region in the United States, a standard would be defined in terms of an acceptable amount of degradation of the land rather than the number of acres mined. A tax schedule could then be constructed based upon this amount of degradation. By taxing land degradation directly, firms would be encouraged to alter their production methods so as to integrate reclamation efforts directly into mining activities. Such a tax would also create incentives for firms to adopt methods of mining that generate smaller amounts of external costs. In areas where flattened mountaintops and exposed high walls (both aesthetic disturbances), landslides, sedimentation, acid water, and compromised wildlife habitat involve damage that more often than not defies complete restoration, the tax on degradation presumably would be very high and would tend to discourage mining there.[18]

None of the states in the Appalachian Coal Basin has defined a zero land degradation standard, but all regulate surface mining under federal guidelines. Reclamation requirements vary some among states, as do the enforcement techniques. This suggests that the elements already exist for an acceptable-standards approach to ensure reclamation in accordance with state or regional needs.

If an acceptable standard for reclamation can be defined, then, specification of workable tax or incentives policy becomes the primary question. The reclamation bond for surface-mining firms is an existing policy tool that is quite close to a tax on land degradation. The amounts of these bonds need to be higher than the actual costs of reclamation if the states want to induce firms to reclaim surface-mined land. If externality costs are an increasing function of the rate of output, the costs of these bonds should be structured accordingly. The bond per acre need not be a uniform charge but could rise with acreage mined.

Tailoring the performance bond system to achieve the standard might also reduce the administrative costs of current regulatory controls. Some of each states' resources presently devoted to the direct inspection and control of surface mining probably could be shifted elsewhere. In fact, when the aim is to achieve an acceptable standard of reclamation, direct controls on production may be redundant. If the tax on land degradation is set high enough, firms would choose to alter their production processes. A small number of inspections of the site should be sufficient to determine whether the mined land satisfied the reclamation standard.

Finally, communities should be free to change the acceptable standard for land degradation at a future time. When heightened concern over land degradation by surface mining is evident to citizens and the policymaker, one would expect a tighter standard to emerge. A change in the standard need not imply an expansion of control over the industry. Upward adjustment in performance bond values should be sufficient to ensure that the rate of degradation is adjusted.

SAFETY IN COAL MINING

In comparison with the literature available on the effects of surface-mine legislation and of the new clean air standards, the literature on the human and economic consequences of the 1969 Federal Coal Mine Health and Safety Act (CMHSA) is skimpy. Considering the resoluteness with which we defend the inviolability of human life, this paucity is surprising. On the other hand, it is well known that business's preoccupation with productivity and profitability often interferes with sound safety procedures. Historically, nowhere has this been more true than in the coal-mining industry, both in the United States and abroad. National Safety Council figures for 1966 put the frequency of disabling injuries in coal mining at 34.2 per million labor hours worked, almost five times the national all-industry rate of 7.2.[19] Among U.S. industries, coal mining was considered the most dangerous to human life. This fact, coupled with the fear that accidents were on the upswing, led Congress to pass the health and safety act of 1969. In no small measure was the passage of the act influenced by the unfortunate 1968 mine disaster in Farmington, West Virginia, in which seventy-eight miners were killed in a gas and dust explosion.

The 1969 act was preceded by two pieces of legislation. The first, the Coal Mines Inspection and Investigation Act of 1941, substantially reduced fatality rates.[20] Between 1940 and 1953, the year the second act went into effect, these rates fell from 1.68 to 0.89, a decline of nearly one-half. The second piece of legislation, the federal Coal Mine Safety Act of 1952, resulted in no decline in fatality rates, even after the act was amended in 1966 to add small mines with fewer than fifteen employees to the inspection lists maintained by the Bureau of Mines. Prior to this amendment, the emphasis of the safety program was on the prevention of major disasters, that is, accidents in large mines. With the 1966 amendments, the focus shifted to the comprehensive promotion of general safety in both large and small mines.

Historically, the U.S. Congress has had to be prodded into the passing of safety legislation. Disasters at Bartley, West Virginia, and Neff, Ohio, led directly to increased public support for the 1941 act, just as the Farmington accident marshalled public support for the 1969 act.

In essence, the 1941 act authorized federal inspections of coal mines but no penalties for violations of safety practices. A federal coal mine safety code was issued in 1946, but it was not until 1952 that the Bureau of Mines was empowered under the Coal Mine Safety Act to close mines determined to be unsafe. Jurisdiction at first extended only to mines with fifteen or more employees engaged in interstate commerce; not until 1969 was the federal government's jurisdiction over safety extended to cover all mines, including surface operations. Moreover, for the first time the issue of safety was redefined to include hazards to coal miners' health attributable to their employment (mostly because of the prevalent pneumoconiosis, or black lung disease), and a new agency was created, the Mining Enforcement and Safety Administration (MESA). Subsequently MESA's name was changed to the Mine Safety and Health Administration (MSHA), and the unit was relocated to the U.S. Department of Labor. Today the agency's responsibilities encompass not only health and safety but education, training, and research as well.

The dramatic decline in average fatality rates over the period 1932-1980 is shown in table 67. The first safety act, enacted in 1941, resulted in a substantial decline in fatalities. The fatality rate declined by 23% over the next ten years, a consequence of numerous recommendations made by federal inspectors, who for the first time were admitted to mining operations. Even though these recommendations were only advisory, not mandatory, they were accepted by most mines and implemented at relatively modest additional costs. The next safety act, passed in 1952, had a much weaker impact on mine safety. Fatality rates over the following sixteen years remained nearly unchanged, declining by only 8%. This was because small mines were not subject to safety regulation at all, even though they experienced the largest number of fatalities.

Another reason for the faint improvements brought about by the 1952 act was neglect by the United Mine Workers of America (UMWA) union. The organization was more concerned with negotiating for broader benefits and higher wages than with safety issues during this period of increased mechanization in mining. Although admittedly the following explanation is speculative, it is also plausible. During the 1950s and 1960s a declining labor force in the mining industry resulted in a marked change in the structure of the UMWA electorate. The share of the total vote held by union retirees as compared with the share of active miners rose sharply. The ratio of beneficiaries in the UMWA Welfare and Retirement Fund to total membership increased from 6.1% in 1950, to 26.8% in 1960, to 46.9% in 1970.[21] It is reasonable to argue that retirees place a different value on workplace safety than active miners and that they have little to gain from enhanced safety. Consequently, the policy orientation of the union leadership with regard to the implicit trade-off between gains in safety and gains in other benefits changed in favor of the latter. To a much larger degree than in earlier years, the leadership in the 1960s had to take into account the preferences of the expanded voting strength of the retired. Although the intent of the sponsors of the 1952 act was most certainly to promote mine safety, it turned out to be weak legislation enacted during a time of weak coal markets and without committed support from the UMWA. The act's failure to improve mine safety is therefore not very surprising.

In contrast, the 1969 act is considered strong legislation followed by prosperous years for the industry. An improved safety record was to be expected, but the dramatic reduction in fatality rates that followed exceeded expectations. Between 1970 and 1980 the rate declined by 54% and it continues to fall in the 1980s. Small mines that heretofore had been exempted from the act now have to abide by federal standards or leave the industry. Many marginally profitable mines were forced to do just that, and the average size of mines gradually increased. The number of small, often unsafe mines has declined significantly.

Accompanying the passage of the 1969 act was also a commitment to greater government spending on mine safety inspection and enforcement. This greatly enhanced the salutary impact of the act. In March 1970, only 327 mine inspectors were employed by the federal government. By 1972 this number had risen to 1,521. Mandatory mine inspections were set at a minimum of four per annum. These inspections cover all aspects of mining: ventilation, roof supports, fire protection, evacuation planning, dust and noise levels, hoisting, loading and haulage, training and certification, and a host of other rules pertaining to health maintenance.

The passage of the health and safety act of 1969 did not reduce greatly the advances in productivity made by the industry during the preceding decades. Immediately following the passage of the act, however, productivity did decline in the major Appalachian coal-producing states. Between 1969 and 1971 it fell at least 21% in Eastern Kentucky and more in the other states;[22] by 1972 the downward trend had stopped, only to be renewed during the years 1973-1976. This decline was occasioned less by the aftereffects of the 1969 act than by the rapid expansion in numbers of coal producers. The sudden surge in the demand for coal following the oil crisis coincided with deep cuts in government-funded heavy construction of roads, bridges, and dams. The cutbacks compelled firms with idle earth-moving equipment to search for alternate business opportunities, which they found in an expanding surface-mine sector in which their equipment was readily employable. The employees of these industries frequently migrated also to the booming coal industry. The addition of these and other new and untrained workers; the opening of new, less-productive mines; the entrance of inexperienced and undercapitalized firms into the industry; and a slowdown in technological progress combined to cause a decline in productivity. By 1977, however, most of these influences had been corrected and productivity levels began to stabilize.

Table 68 shows a comparison of safety in underground and surface mining for the period 1930-1978 in the Appalachian states. Two researchers using federal statistics measured tons of coal produced per fatality during each of ten time periods. The trend toward increased safety as reflected in an increasing number of tons mined per fatality is unmistakable. The only blemish on the safety record is the 1975-1978 period in surface mining. Just as productivity declined with the influx of inexperienced operators after the oil crisis, the fatality rate increased temporarily, until the late 1970s.

Of some concern is that while the CMHSA has reduced fatality rates, nonfatal accidents may be on the rise. This raises a serious issue, namely how to evaluate the effectiveness of the act—in terms of fatalities only, or more broadly in terms of its contribution to overall safety, measured in terms of fatal and nonfatal accidents combined.

Finally, there is the question of the effects of the use of safety equipment, procedures, and experience on industry competition. Small firms with low annual output are unable to take advantage of economies of scale and to distribute the additional costs of safety over a large number of tons of output. Consequently, their safety costs, which largely are fixed and unrelated to mine size, are higher per unit produced than for large firms. Such high costs can force small firms out of the industry and reduce competition. The extent to which this might have happened already is unclear. However, the evidence suggests that ample competition remains in the regional markets.

Most of the costs of implementing the requirements of the CMHSA are passed on to the consumers of electricity. Contractual agreements between coal producers and electric utilities generally stipulate that the additional costs that arise from meeting new safety, environmental, or tax legislation will be passed on to the buyer of coal. The electric utilities in turn pass their costs to the consumer in the form of fuel adjustment differentials. Only in the case of the rather small amount of coal traded in the spot market are some of the additional safety costs borne by the mine operator.

In conclusion, the Federal Coal Mine Health and Safety Act has eliminated at least a portion of the lag by which the coal industry followed other industries in protecting its workers. The act resulted in a productivity decline and hence increased costs at least during the years 1970-1973.[23] Beyond these years further declines in productivity are attributable to factors other than tightened safety regulations.

VII

COAL TRANSPORT
AND EXPORT

To virtually all coal producers, the cost of transporting their product to market is of vital importance. Transport can involve one or several types of carriers.

A modest amount of coal is moved from mine mouth to rail tipple by conveyor belt. These conveyors require a substantial initial capital investment, involve maintenance costs, and are relatively inflexible. They can be realigned and extended, but their reach is limited. Generally motor carriers are more appealing for the movement of coal to the tipple. These carriers fall into three classes. Owner-operator carriers are individuals who own one or more trucks; usually fiercely competitive, these individuals most often move coal for independent, generally small operators. A second carrier class includes fleet-owned trucks, vehicles owned by medium-sized and large mines or by large consumers of coal and serving the interests of the truck owners. The third class is the common carriers, who serve extended geographic regions; these transport a variety of commodities, but only occasionally do they carry coal.

In Appalachia, the dominant carrier of coal is the independent truck owner. Truck size and type are often prescribed by the nature of the roads the owner-operator travels, their steepness and state of repair, and by state and federal weight limits on roads. In areas where a high volume of coal is transported, shippers have the option of choosing among competing truck operators. In these areas service is generally better and transport costs are lower.

Once the coal reaches a rail link, it is typically unloaded into a tipple or preparation plant where it is washed. From there the coal

is loaded into rail cars for long-distance transport. Tipples and preparation plants are costly to erect and to maintain. They may be owned
by coal companies or by individuals, but rarely are they the property
of the railroads. Because of the size of the initial capital investment,
owners of preparation plants prefer to enter into long-term contracts
with high-volume mining operations. It is often the case that out-of-
the-way mining operators have to truck their coal considerable
distances in order to reach a tipple and rail point. In such cases, the
costs of transporting a ton of coal may exceed substantially the overall
average transport cost of two dollars per ton. In fact, the truck
transport of coal follows no prescribed cost guidelines and is subject
to a very wide range of special agreements and circumstances.

RAIL TRANSPORT

Railroad transportation is vital to coal producers. More than two-
thirds of the coal moved each year travels either largely or entirely
by rail, and the railroads have been an integral part of the coal industry since its inception. Table 69 shows the amount of coal moved
by rail for the period 1970-1983 and the share of coal traffic in the
total freight handled by the railroads. The data illustrate that no
dramatic change has occurred in the mode of transporting coal in recent years. The railroads handle approximately the same share of all
coal transported as in 1978. Between 1970 and 1981, however, coal
has increased significantly its share in the total tons of freight moved
by the railroads, from 27% to 36%. The remainder of coal is
transported by barge or truck. In 1983, 113 million tons, or 16% of
all coal shipped to U.S. customers, were moved by barge; 100 tons,
or 14%, by truck; and 78 million tons, or 11%, by conveyor belt or
slurry pipeline.[1] Except for a few instances in which electric utilities
or steel-producing plants are located close to mining operations,
railroads are the principal carriers, even over distances as short as 50
miles.

For most buyers of coal, the principal criterion in the decision
to make purchases is the delivered price of coal, which includes the
costs of transport. Changes in transport costs can materially affect
the demand for a particular operator's coal because such changes alter
the delivered price to the buyer. Both mine operator and consumer
have a common interest in keeping transport costs to a minimum
and therefore in railroad technology and pricing policies.

In recent years the most important innovation in rail coal
transport has been the introduction of unit trains. These operate much

like a shuttle service between a point of origin and a destination, usually on a fairly precise schedule. First put into operation by the Southern Railroad in Alabama in 1960, these trains have expanded rapidly and steadily in number and use. In 1980 approximately one-half of the coal transported by the railroads was moved in unit trains.

High productivity and lower operating costs make unit trains very attractive. Unit trains are composed of dedicated equipment that is regularly scheduled between two points and has rapid loading and unloading capabilities. They eliminate the need for en route switching. These trains are particularly favored by large producers and consumers who have the most to gain from their use, but the initial capital investment in rails, equipment, and loading terminals can be high.

Characteristically, small mine operators have a less favorable view of the benefits of unit trains. They are unable to capture the cost benefits associated with unit train shipments and therefore risk losing markets in which their coal transported in single cars is more expensive. The same is true of small-volume consumers of coal who also are unable to participate in the cost savings of unit train transport. Large-volume consumers often are unwilling to pay the higher single-car rates charged by the railroads to small producers. Consequently, they frequently refuse to deal with small mining operators who ship coal in single cars.

Some railroads, and some large buyers of coal such as the Tennessee Valley Authority (TVA), deal with this situation by establishing gathering points for small producers. The TVA also sets aside a certain portion of its fuel needs to be met by small producers only. Overall, small coal producers continue to be concerned about the eventual effects of deregulation of the railroads. Their fears center on the prospects of being abandoned by the railroads either through the closing of low-traffic rail lines, or by being confronted by high rail rates that may price their coal out of the market.

The maximum load-carrying capacity of modern rail cars is 100 tons. Because the empty weight of the car is approximately 60,000 pounds, a fully loaded car can weigh 131 tons. This load, distributed over four axles, represents a considerable amount of weight placed on what is often a fragile, ill-maintained roadbed. At the height of the recent coal boom, derailings occurred with unacceptable frequency and at considerable cost to the railroads. The most widely used of the 300,000 U.S. rail cars are of the hopper type. Sloping plate bottoms and gates allow quick discharge and short loading-unloading times. Most unit trains carry between 7,000 and 10,000 tons, and in some instances between 10,000 and 15,000 tons. Fewer than 10% of the rail cars are owned by shippers or consumers; the remainder are property of the railroads.

Prior to 1974 coal transport rates were set by the railroads mostly with an eye toward the delivered price of coal substitutes such as oil and natural gas. Following the Arab oil boycott and the middle-1970s gas shortages, the railroads began to raise rates steadily. In the late 1970s the Interstate Commerce Commission (ICC) proposed a cost-based rate structure that distributed fixed costs based upon tonnage and ton-miles, including the current costs of capital. In 1981 the ICC revised this procedure in favor of pricing in accordance with varying market demand elasticities.[2] In short, this scheme specified that the markup above attributable costs would be greatest where transport alternatives were smallest and lowest where competition was most intense. According to ICC reasoning, this would maximize railroad revenues and generate the most efficient use of equipment.

Most of the contemporary regulations governing rate-setting procedures of rail transport are an outgrowth of two pieces of legislation: the Railroad Revitalization and Regulatory Reform Act of 1976 (the so-called 4-R act), and the Staggers Rail Act of 1980.

As the first attempt at railroad deregulation, the 4-R act addressed principally the issue of market dominance of carriers. The act's provisions also instruct the ICC to consider the question of "revenue adequacy" in its rate approval policies.

The Staggers act is specifically designed to cushion the long-term deteriorating financial conditions of the railroads. It assumes that in the rail transport industry competition is the rule, concentration the exception. Therefore, it encourages the railroads and coal shippers to enter into independent, long-term transport contracts for greater stability and improved service. In the past, the ICC had been required to approve rate increases and tended to set rail rates that covered at least the variable operating costs and made some contribution to fixed costs. Requests for more generous rate increases often were denied. Under the Staggers act carriers are authorized to raise freight rates in accordance with inflationary cost increases with only minimal ICC review. The net result of these new rules has been to shift the burden of proof in disputes over rail rates to shippers in that the act requires them to substantiate claims that rates are unreasonably high. Predictably, coal producers and shippers are dissatisfied with the Staggers act and its implementation. The railroads, on the other hand, are generally pleased.

In particular, the issue of captive shippers and the rates charged them under the Staggers act needs further study. These are coal producers, such as coal exporters, who have only one way of shipping coal to its destination; the producers of Virginia coal who export it to Europe are an example. Shipping this coal to any port other than Norfolk, Virginia, would greatly raise its delivered price. For all intents and purposes, these shippers are captives of the railroad serving Norfolk.

At this writing it is too early to assess with any degree of certainty the current economic impact of the Staggers act. Nor is it possible to predict its future impact because modifications in ICC regulations continue to be made and the courts are still being asked to adjudicate disputes. The Association of American Railroads supports with enthusiasm the market freedom gained from the act's passage and the positive impact this is having on the industry.[3] The National Coal Association, on the other hand, continues to challenge many of the act's provisions and is actively lobbying Congress to amend the act and close its loopholes. On balance, it is clear that the act has led to higher rail rates, particularly in captive markets. In many coal-producing regions, shippers of coal have little if any choice in the method of transporting their coal to markets. Only if they are willing to incur the additional costs of long-distance truck hauling to, for example, another rail line, can it be said that some shippers have a choice. In many instances this alternative is infeasible because the road distances are simply too great. The delivered price of coal transported thus might then be even higher and exceed that of the competitors' product.

It is clear that the railroads have a problem of revenue adequacy. Adequate revenues are needed to cover the costs of track maintenance, car and locomotive ownership, interest, and taxes. Years of neglect followed by a sudden surge in the demand for coal transport capacity has resulted in numerous derailings, low productivity, and high transport costs. Redistributing these costs to make them reflect more accurately their true origins is one of the goals of the Staggers act. From an allocative standpoint, this redistribution would create greater efficiency. In the process, some mining operators and shippers, in particular the smaller ones, would confront higher transport costs, which in many instances would contribute to their inability to compete.

Historically, transport charges have represented a significant but declining fraction of the delivered price of coal. In particular this was the case in the years after the boycott. But with falling coal prices and modestly rising rail rates, this trend has reversed itself lately.

The two economic recessions of the first four years of the 1980s, with perhaps a third one on the horizon for 1985-1986, clouds the coal transport picture further. The railroads, chronically short of capital, tend to postpone the purchase of new, more efficient capital equipment and the upgrading of the track system until a significant increase in demand compels them to do so. Because the factors that influence the demand for rail transport are likely to face demand problems themselves, it is unlikely that in the foreseeable future this demand will be stimulated greatly.

OTHER TRANSPORT MODES

Barges provide for some shippers and consumers an alternative mode of transport. The two principal inland waterways are the Ohio River and the Monongahela River. Over 50% of the coal shipped by water moves on these two rivers. The water carriers are largely unregulated with regard to rate making. Safety and navigation aids, such as dams and locks, are maintained by the U.S. Army Corps of Engineers at no cost to users, so in a sense these water carriers are subsidized. Because most coal transported by barges moves downstream, there are no grades that must be negotiated. Rates are based largely on distance (see table 70).

The barge industry is highly competitive, with approximately 1,800 firms competing for business. Contract agreements between shipper and barge owner are common, a fact that makes this link in the coal supply chain attractive to the shipper. Long-term contracts permit the barge owner to modernize, update, and plan equipment use and pass on at least a portion of the resultant cost saving to the shipper in the form of lower per-ton transport rates. Some electric utilities own their own fleet of barges and river terminals and thereby realize substantial cost savings.

The issue of the slurry pipeline transport of coal suffered a disappointing defeat at the hands of Congress in 1983. Under considerable pressure from an apprehensive railroad lobby, Congress refused to pass legislation that would have given slurry pipeline companies the right of eminent domain on railroad property. This defeat caused, at least for the present, the abandonment of several proposed pipeline projects. Among them were the Coalstream project connecting Appalachian shippers to Florida, Georgia, and export ports, and the Transco-Vepco pipeline in Virginia, which could have resulted in savings of $5.87 to $8.66 per ton over current rail rates.[4]

PORT CAPACITY AND COAL EXPORTS

When coal moves from an Appalachian tipple to a port of departure, it will be transported in one of several ways. The so-called line-haul can be by truck, barge, or rail, or by a combination of these methods. The type of line-haul used and the port of departure accessed depends on the location of the shipper.

The cost of transporting coal by barge to a port of departure can vary over a wide range. It is influenced by the degree of market competition among carriers, the length of the haul, the volume of coal hauled, and the ownership of the carrier. Table 70 shows some representative barge rates and highlights their great variation.

A shipper not located within a few miles of a waterway will probably transport coal during the first leg of the journey by rail to an inland waterway port. If it is more attractive to ship the coal to a port of departure on the east coast, it will go by rail. As a rule, a shipper is served by only one rail carrier and will therefore access the port or ports served by that carrier. For example, it is unlikely that an Eastern Kentucky shipper from Pike County will find it attractive to send coal north to the nearest Conrail spur destined for Philadelphia. Most likely, it will be less costly to send coal via the Chessie System to Hampton Roads, Virginia. In other words, the geographic locations of the mining operation and the shipper dictate the appropriate port of departure. Few, if any, competitive alternatives are available to a shipper from among the various rail carriers serving the coal industry in general.

The U.S. seaports are the final link in the inland transport of export coal. The amazing and unexpected export surge in coal from 1979 through 1982 created a serious short-term scarcity of loading capacity. Now that the export demand for coal has ebbed significantly, so has the interest in improving, expanding, and deepening the nation's port capacity to handle more export coal.

Historically, the U.S. export of bituminous coal has been fairly constant, and port capacity has been adequate to handle the volume. However, the dramatic 85% rise in 1979 exports and further gains in 1980 and 1981 underscored the inflexibility and inadequacies of U.S. ports to handle efficiently large volumes of goods, including minerals. In those years the questions not only of the adequacy of rail transport but also of the inability of port facilities to handle 100 million and more tons of export coal per year were being raised with increasing urgency. Of immediate concern were the shortage of coal storage space, of sorting, blending, and loading facilities, and of escalating demurrage costs due to long delays in port. The demurrage costs, which run from $15,000 to $20,000 per day, can range from $450,000 to $1,000,000 per ship during thirty- to fifty-day waiting periods for access to a loading berth. The question of who is to bear the burden of these costs has not been resolved satisfactorily. In contrast, the ability of the U.S. coal industry to produce additional export tonnage has remained unchallenged. Most analysts agree that three years ago, very much like today, the excess annual capacity to mine coal stood at approximately 100 million to 150 million tons. Clearly, export booms are limited more by the shipping industry than by the coal industry.

The unexpected export surge of several years ago arose from two principle sources: (1) the inability of Poland to meet its external coal sales commitments due to political unrest, and (2) labor union strife in Australia. In addition, spiraling cost increases for diesel fuel made the transport of coal from Australia and South Africa to Europe increasingly more expensive. Responding to an urgent European and Japanese search for alternative sources of coal, the United States was able to capitalize on its competitors' problems and substantially increase its spot market export sales.

Even though the delivered price of U.S. coal in Europe is higher than the prices of other major coal exporters, coal importers often are concerned with factors other than price. Among these, security and reliability of coal supply rank very high. In 1980, for example, the delivered price of U.S. coal in the Netherlands was 13% higher than that of the next most expensive supplier and 33% higher than the third most expensive. In Japan, U.S. coal costs 35% more than Australian coal. Still, in both instances U.S. exports surged, driven by considerations other than price.

Nearly all U.S. coal exported overseas is handled by seven ports, one of which—Hampton Roads, Virginia—handled 44% of all U.S. coal exported in 1984 (see table 71). In 1982, with five active piers, Hampton Roads had excess capacity of 25%. The second largest port for overseas exports on the East Coast was Baltimore, which in 1984 handled about 9% of total U.S. exports. In 1984, however, Mobile, Alabama, put into use its vastly expanded port facilities. Export capacity now exceeds 30 million tons per year and by 1990 it is scheduled to double. If anticipated deepening of the draft to 55 feet does take place, colliers will be able to load from 120 thousand to 130 thousand tons, a very efficient way to transport coal. This presages the arrival of an enhanced competitive environment for coal port facilities and for the rail and barge transport that service them. Other overseas ports of considerable size are New Orleans/Baton Rouge and Philadelphia.

None of the U.S. coal-exporting ports approaches the maximum drafts (harbor depths) of Australia, Canada, and South Africa. The Hampton Roads draft is 42 feet and the New Orleans and Mobile drafts are 40 feet. In contrast, the two deepest Canadian ports, Prince Rupert and Vancouver, have 65-foot drafts; Australia's and South Africa's port depths range between 53 and 56 feet. During 1980 and 1981 a flurry of plans was hatched to expand port capacity by deepening channels to be able to accommodate 100,000 to 150,000 deadweight-ton (that is, empty-weight) supercolliers. Before most of these plans could be implemented, however, the coal export boom began to show signs of abating. Because of the weakened state of the coal industry today, it is unlikely that many, if any, of the ambitious expansion and port-deepening plans will be fully implemented in the near future. At this time, U.S. coal-handling ports can accommodate only ships up to an empty weight of 80,000 tons. All European and Japanese ports can receive ships with a draft of 55-79 feet. Thus, if the large supercollier class of oceangoing ships is built in the future and becomes the common mode of ocean transport, U.S. shippers and U.S. ports will have to operate at a competitive disadvantage. It is feasible, for example, that Australia's and South Africa's competitive disadvantage in distance will be offset by their ability to load coal into supercolliers and thereby reduce its delivered price. Because of economies of scale it is estimated that a 100,000-ton ship can transport coal from the U.S. East Coast to northern Europe at $1.75-per-ton capital costs and $1.27-per-ton operating costs. The comparable numbers for a 60,000-ton ship are $2.37 and $1.83, all in 1980 dollars.[5] The cost savings are substantial.

There exist, of course, several alternatives to coal port deepening. These have been underscored recently because of a reluctance of public authorities to finance port deepening and because of environmental concerns. Designs for shallow draft, wide-beam ships in the 144,000-ton class are being studied. Topping-off operations by ships or barges offshore also are being considered, as are slurry-pipeline feeds to vessels anchored offshore. These alternative loading methods are designed to circumvent the requirements of sizable capital expenditures on port improvements. Whether they are a long-term solution to the disadvantages of shallow-draft ports is uncertain at this time.

In retrospect, the judgment of U.S. decision makers not to be stampeded by a short-lived coal export boom into the wholesale expansion of port capacity appears to have been a wise one. If all construction, expansion, and modernization planned in 1980 had been implemented, current port capacity to export coal would be in excess of 200 million tons annually, and much more than one-half of it would have been idle in 1983 and 1984. One is reminded of the frantic shipbuilding boom years of the 1960s when transporting another fuel—oil—in supertankers became fashionable. Country after country geared up to expand its capacity to build ships: Japan, Sweden, Germany, France, Greece, England, Italy, and others. The United States did not. By the 1970s, many of the new shipyards were in financial difficulty and were able to extend their life only with government subsidy. Others closed their doors or converted to the construction of bridges, barges, and other equipment. Had port capacity in the United States been expanded with equal haste, it is quite likely that many ports would be in a state of serious underutilization today. It would seem that the wisest approach to port deepening and modernization would be a joint cooperative approach involving coal companies, the railroads, municipalities, the federal government, and perhaps foreign buyers. This cooperative approach should have as its objective not regulation, but cooperation in order to improve the efficiency and cost of the ocean shipment of coal.

VIII

PROSPECTS FOR APPALACHIAN COAL

It is sometimes said that predicting the future resembles a game of chance. Accurate prediction does, of course, involve something more than fortuity. This chapter attempts to meld the lessons of the past with educated intuition and to arrive at responsible conclusions as to what can be expected in the coal industry.

The first assumption made by economists who attempt predictions is that human nature and behavior are governed by basic laws of economics that remain relatively constant over time. That is, we tend to respond individually and collectively as our predecessors did. Implied when one predicts the future is the belief that it can be manipulated somewhat according to the options that we include or exclude from the forecast. The seriousness of the endeavor becomes apparent when the task is viewed in this light. The predictions here are offered with respect for an industry that has had a long and honorable struggle in the past and with the hope that future energy policy, whether initiated by public authorities, the industry, or consumers, will continue to benefit the individual as a social being.

Fuel Alternatives to Coal

For coal to prosper in the future as an energy resource, it will have to compete successfully in the market with other fossil fuels—oil and natural gas—and to an extent with nuclear power. Current technology allows the use of four major energy resources: coal, petroleum, natural gas, and uranium 235. In the future, other fuels such as shale oil, uranium 238, and solar energy may join the list.

Of the fossil fuels, only petroleum, coal, and natural gas are obtainable at present in sizable amounts. A fourth, shale oil, is in an uncertain state of development. Commercial recovery of this resource does not appear imminent.

The enormity of U.S. coal reserves is overwhelming, particularly when contrasted with the nation's gas and oil deposits. Table 72 shows the magnitude and distribution of world fossil fuel reserves. Only the coal deposits of the U.S.S.R. approach those of the United States; China's reserves run a distant third. Oil and gas reserves, in contrast, are relatively small. There are many nations in the world whose petroleum and natural gas deposits greatly exceed those of the United States and for whom these deposits represent the principal, if not the only, source of foreign exchange earnings. China, Mexico, Venezuela, Saudi Arabia, the U.S.S.R., Iran, and Kuwait are cases in point. Nearly all of these nations are in the process of development (or redevelopment), and earning convertible exchange is an extremely important goal in their plans. In short, the nations pay for their imports of Iowa corn and grain and Ohio machinery by exporting oil and natural gas to Germany, France, Italy, and other western nations.

The Chinese experience illustrates the importance of international trade to development. A large part of its foreign currency earnings comes from the export of labor-intensive goods, such as textiles, shoes, and other commodities. The nation is just now beginning to develop its extensive fossil fuel reserves and it needs to import from the West oil drilling, refining, and pipeline equipment. To accomplish its goal it has to earn convertible exchange. Because of its ability to export its textiles, China can be expected to compete more and more extensively in the world oil market.

The U.S.S.R.'s dependency on its mineral resources may be surprising to some. Nevertheless, its only significant sources of foreign exchange are the sale of oil, gas, and gold. In 1982 about one-half of its hard currency receipts came from oil and gas exports.[1] Estimates for the first half of 1984 show that Soviet oil output, the largest in the world, averaged 12.3 million barrels per day, 22% of a global output of 55 million barrels per day. Middle Eastern production averaged only 12 million barrels per day, and China, although it raised its production of oil in 1983 by 7%, still produced an average of only 2.2 million barrels per day.

That the world is generously endowed with petroleum reserves is clearly illustrated by the statistics in table 72. The data represent conservative estimates; because deposits are poorly charted and measured in developing nations such as China, Mexico, and

Venezuela, it is generally believed that actual reserves greatly exceed estimated reserves. Reserve estimates for the developed nations—the United States, Canada, and Norway, for example—are considered more reliable and representative of actual deposits.

Historically, the welfare of the coal industry has closely paralleled fluctuations in the price of petroleum and, therefore, the welfare of the oil industry. The price of coal moves in the same direction as the price of petroleum, albeit at a lower level and with a time lag. This so-called price tracking occurs because when the price of oil rises, coal becomes an increasingly attractive substitute. Market demand increases tend to raise the price of coal; abatement of demand tends to reduce it.

A vivid illustration of the parallels between the two industries is provided by the coal-boom years of 1974-1975 and 1980. Prior to the Arab oil embargo, price increases were largely cost induced by such events as the enforcement of new mine safety legislation and UMWA wage settlements. After the unexpected Arab actions, however, the price of coal, in particular the spot-market price, rose because embargo restrictions reduced sharply the available oil supplies and because the price of imported oil had increased fivefold. Uncertainties concerning the future availability of adequate supplies of imported oil, coupled with sharply higher prices, generated a strong upward shift in demand for its most suitable substitute: coal. The price of coal rose dramatically. A similar situation developed following the political upheaval in Iran in 1979 and 1980, but on a more modest scale. Since then, however, the major oil-importing nations have reduced their reliance on Middle Eastern oil, undertaken successful energy conservation measures, built up strategic oil reserves, and expanded their use of other energy resources.[2] The result is that future interruptions in world oil supplies, as long as they are of modest proportions, are unlikely to have a serious impact on the economies of the developed nations.

Future changes in the price of petroleum and their impact on the Appalachian coal industry are difficult to forecast. Recent events seem to support the view that in the long run real petroleum prices will decline modestly under competitive supply pressures. Mexico and China are entering the market as major suppliers of oil. The U.S.S.R. will unquestionably continue to sell substantial quantities of oil in the world market, and the monopoly power of the Organization of Petroleum Exporting Countries (OPEC) is becoming seriously limited.[3] These events are not particularly surprising. Only the fact that it took nearly ten years for them to happen is remarkable. On economic grounds, cartels are traditionally unstable. If one large or several small members break away from the cartel's policies, a stampede for the exit can easily develop. Historically, cartels are formed to acquire monopoly profits by interfering with the operation of the market. But if output can be expanded and sold at production costs that are below the market price, the threat to monopoly profits from competitive pricing becomes very real. This has happened and will continue to happen in the case of OPEC, in which several member nations, notably Nigeria, Iran, and Iraq, have had an insatiable need for convertible currency. Nigeria's needs come from extremely ambitious and perhaps unwise development plans requiring large amounts of western imports. For Iran and Iraq, the importing of armaments and the need for rebuilding in the war-torn countries have generated an acute need for foreign exchange to finance the western imports. These countries have found it difficult to ignore the potential rewards of circumventing cartel production quotas and prices. They have frequently exceeded official production quotas and sold oil at discounted prices in the spot market. By the end of 1985, spot prices had fallen to long-term lows, well below earlier contract levels.

The coal industry is very vulnerable to the potential of falling oil prices. Coal and oil are competitors in world energy markets. In the post-World War II period, oil has systematically squeezed coal out of the electric utility, industrial, and manufacturing markets, except in markets located very close to the coal-mining states. Oil is clean, versatile, easy to handle, and has a higher energy content per unit of weight. When compared with those of coal, petroleum's virtues are many.

Coal also has two other competitors, nuclear power and natural gas, both of which have expanded strongly their market shares. Nuclear power has seized a significant portion of coal's most traditional market, the electric utility power generation market. It is unlikely, however, that it will expand this share further in the next decade or two. Safety and environmental concerns, along with dramatic escalation in construction costs, have sharply curtailed plans for additional nuclear capacity in the future. At the core of nuclear power's unpopularity lies the general public's perception of nuclear safety. If it is accepted that the majority of voters are economically risk adverse, it is not surprising that they are unwilling to accept the risk of a nuclear accident, no matter how small the probability of its occurrence. The public correctly perceives that a far larger number of people would be affected by a nuclear power plant accident than by a disaster at a conventional power plant. Virtually the entire risk of the latter is borne by utility employees and some of the residents in the contiguous area. In a nuclear plant accident, however, the risk would be spread to a large number of people working or living within hundreds of miles of the facility.

The risk to life of using nuclear generating plants is often considered to be significantly higher than the risk to life of using fossil-fuel-fired generating plants. Professor Kazmer has suggested, however, that comparisons of risk in the two types of facilities are often incomplete.[4] The correct comparison is between the probability of morbidity and moribundity from all aspects of nuclear power generation, including the likelihood of accident in uranium-mining as well as the probability of a power-generating accident and its consequences, and the probability of morbidity and moribundity from all aspects of fossil-fuel power generation.[5] The latter would consider accidents, disablement, and death in the mine during coal transport and coal combustion and during waste disposal. The broader comparison would result in more meaningful decisions about the relative merits of nuclear-fuel and fossil-fuel power generation.

As an energy source, natural gas, which entered the American industrial markets in the 1940s and 1950s and began to be used somewhat later in European areas, following the Dutch discoveries at Groningen, seems destined for curtailment in the United States. In Europe, on the other hand, its use has grown sharply, mostly at the expense of coal and oil use and as a result of Soviet and Western European commercial agreements.

In the United States, government control over natural gas traded in the interstate market started thirty-five years ago. In 1950 it was feared that natural gas reserves would soon be depleted and that consumers would be required to pay economic rents to the handful of suppliers able to continue production. A price ceiling on gas sold between states was established to protect consumers. In response, suppliers increasingly began to shun the interstate markets and confined their sales to unregulated intrastate markets. The rising oil prices of the middle 1970s, which increased sharply the demand for natural gas as a substitute for oil, and the accompanying decline in the availability of interstate gas created a troublesome natural gas shortage. Following intense lobbying by special-interest groups, Congress passed in 1978 the Natural Gas Policy Act (NGPA). Unfortunately, the act's provisions have failed to inspire confidence in its ability to solve the problems. Old gas traded between states remains price controlled. Gas discovered and mined after 1977 can be sold at higher prices. Gas extracted from greater than 15,000 feet below ground is totally deregulated. As one would expect, the effect of the act has been to encourage extensive and expensive deep drilling.

The future of natural gas is difficult to predict. Falling oil prices, coupled with uncertain economic conditions and excess oil-production capacity, make an increase in natural gas prices unlikely in the near future. The use of natural gas in electric power generation is gradually being phased out, and the most promising potential for the expanded use of natural gas seems to be in the industrial and commercial sectors. The pace of this expansion is likely to be modest and will depend to a large degree on future economic conditions.

The 1985 changes in natural gas regulation will not affect greatly the demand for coal. Natural gas prices, too, are tied to world oil prices, and these are not rising but falling. Consequently, the coal industry cannot look to decreases in the demand for other fossil fuels as a possible source of expanded demand for its own product as long as the real prices of substitute fuels are declining. In the long run, however, as natural gas reserves dwindle and extraction costs rise, equilibrium prices for natural gas, unlike those for coal, will certainly rise.[6] If other things remain unchanged, the demand for coal as a substitute for natural gas in electricity generation and industrial and commercial processes can therefore be expected to rise as well.

As discussed earlier in this book, in the United States and most of the developed West the principal user of coal is the electric-power-producing industry. Economic health of the coal industry hinges on growth in electric power demand and the extent to which this demand is satisfied by coal-fired generating capacity. This capacity can grow either through the addition of entirely coal-fired facilities or through the conversion of existing facilities now using other fuels. The projected growth rates of electric power generation and of coal's share as a fuel in electricity generation in North America and the Pacific and European countries are shown in table 73, which lists the most recent International Energy Agency projections. The prognosis is remarkably steady although modest growth for North American and European power generation is foreseen. Average annual growth of 2.6% seems consistent with the fact that rising energy prices for the past ten years, coupled with long-run conservation efforts and a reduced growth rate in the world economies, have moderated the historically high growth rates of electricity demand in Europe and North America. The Pacific region, in contrast, is projected to experience a decline in the average annual growth in power generation, although coal as an energy fuel is expected to increase substantially its share of the market. The presumption is that despite the growth of nuclear power production capacity, governments in the Pacific rim nations plan to lessen their dependence on oil by diversifying energy resource use. They plan to do so by actively encouraging the construction of coal-fired plants. This decision has been based in part on political considerations and in part on the proximity of Australian and Chinese coal resources. The decision is also motivated by what is perceived as the lower total cost of producing electric power with coal. An interesting cost comparison prepared by the International Energy Agency shows that under the assumptions of a 15% interest rate and of a ten-year lead time in the construction of a nuclear plant, coal would be the least costly fuel to use in the making of electric power. Table 74 shows cost comparisons for oil, coal, and nuclear power. In the United States, the total costs of generating electricity are lowest for coal. Because of the lengthy delays and cost escalations common in nuclear power plant construction, capital costs are very high in such plants. Consequently, the cost of electricity from such plants appears to be not much different from that of oil-fired facilities. According to these estimates, U.S. electricity from coal is about one-third less costly than oil- or nuclear-based electricity. In Europe and Japan, coal-generated electricity is considerably more costly because the delivered price of coal there is much higher. Also of interest is the difference in the capital investment costs of oil, nuclear power, and coal. At a 15% market rate of interest, capital costs per kilowatt of electricity are approximately three times higher for a nuclear facility than for an oil-based facility and twice as high for the nuclear plant as for a coal-based plant. In a capital-short world, the higher investment costs for nuclear and coal plants are a powerful deterrent to their construction. Not surprising is the fact that, if other things are equal, electric utilities often opt for the less-capital-intensive oil-fired facilities when deciding upon new power plants.

NEW TECHNOLOGY IN COAL USE

Some analysts point out that under current technology coal is the fuel best situated to meet the anticipated energy requirements of the long run. They are convinced that coal is the fuel of the future, even to the point of asserting that it will be used to produce synthetic oil and gas, two of coal's principal competitors. Other analysts are less sanguine about coal's future, preferring instead to condition predictions about future use on the development of new uses of coal.

The major factor that continues to constrain the expanded use of coal as an energy source is not the cost of burning it, but the cost of burning it in an acceptable way. Two recent developments in boiler-fuel technology may in the future remove the constraint on greater coal use as a boiler fuel.

Coal-water slurry as a boiler fuel. Because interest rates are likely to stay high, and because demand for new generating capacity will remain weak, public utilities are expected to be very cautious about adding new power plants. In contrast with their behavior in the 1970s, they are shying away from large, expensive, and sophisticated units and looking toward smaller, simpler, and more flexible plants. They are also examining the feasibility of converting existing oil-fired boilers to use coal-water slurry (CWS).

It is simpler and less costly to use this slurry than to use oil. Composed of about 70%–75% crushed coal, 24%–29% water, and 1% chemical additives, the slurry has the approximate consistency of latex paint and burns much like oil. When expansion in power generation plant capacity resumes sometime in the future, probably in the 1990s, the alternatives for utilities will be to acquire new coal-fired boilers or to convert existing oil-fired ones to use coal-water-slurry fuel. High initial investment costs make the construction of new coal-fired plants unlikely.

Coal-water slurry is stored and transported in much the same manner as oil: in tanks on railroads, ships, or trucks, or by pipeline. Prices compare favorably with oil prices: oil costs approximately $4.60 per million Btu produced, and CWS costs from $2.00 to $3.50 for an equivalent number of Btu. There are, however, other costs that must be considered, such as those of transportation and plant conversion to enable the use of slurry. In final analysis, the advisability of using CWS is conditioned by location, the price of oil, and the enforcement of emission controls. The basic question is whether the Environmental Protection Agency will (1) adhere to an opinion in which it was held that oil boiler conversions to CWS are not subject to federal new source performance standards, or, (2) as a 1983 opinion required, regard the issue a matter of state jurisdiction. At the state level, regulations and enforcement vary greatly, and utilities are prone to postpone decisions until the requirements are more clearly defined.

Fluidized-bed combustion. The national concern with clean air and the reduction of power plant emissions has led researchers to examine more carefully the merits of fluidized-bed combustion (FBC). This process of generating electricity is inherently clean because it emits little SO_2 and NO_x. Experiments to determine the economic and technical feasibility of FBC are now under way at several locations, in particular at the Duke Power Company in North Carolina and at the TVA's Shawnee plant in Paducah, Kentucky.

The FBC process involves the burning of finely crushed coal fed continuously into a turbulent (fluidized) mixture of granular limestone. Both are suspended in a special furnace by pressurized air flows. The limestone granules capture 90% or more of the SO_2 produced by the sulfur in the coal. The combustion temperature is 1,500° F, about one-half the temperature required in traditional boilers. This technique not only allows the burning of low-quality fuels but also prevents slagging and the formation of NO_x. For the high-sulfur coal-producing regions of the nation, this technology is particularly attractive because it eliminates the need for costly flue-gas desulfurization systems. Thus FBC may be the method used by electric utilities to convert existing boilers using pulverized coal. It may prove to be a considerably less expensive method than to install flue-gas desulfurization equipment, which has high investment and maintenance costs. The next generation of coal plants will surely use cleaner combustion technology, including FBC. Some countries already have FBC units operating at various scales. Russia and China have relatively simple units, while the most sophisticated are in use in Scandinavia. A considerable amount of research on FBC is continuing all over the world.

THE ISSUE OF ENERGY POLICY

The events that surround economic decisions on energy are nearly always critical to their outcome, and yet they are unpredictable. Even if the United States had a comprehensive energy policy, demographic, environmental, economic, and political events at home and abroad would exert constant pressure for change. In fact, the United States does not have such a policy. Instead we have individual policies, rarely coordinated, that influence national security, the environment, natural gas extraction, mining safety, and energy transport, to name a few areas. It appears that energy policy does not rank high on the priority list of either the administration or Congress. This fact itself, however, is reflective of the mood and intent of Congress—that is, not to impose a comprehensive independent energy policy on the nation. In not legislating such a policy, a statement on the need for an energy policy is made indeed.

PROSPECTS FOR COAL CONSUMPTION

Prospects of the coal industry, especially of the Appalachian coal sector, will be determined to a large extent by government policy at home and abroad. As desirable as it may be for efficiency and the maximization of social welfare to pursue a policy of free trade in energy resources, the politics of national interest will continue to dictate otherwise. For the remainder of the decade, domestic coal consumption and coal exports are likely to grow only modestly, consistent with historic trends. In the relatively stable energy environment of today, and in the absence of traumatic shocks, coal production, price, and profit levels are not expected to diverge greatly from trends established over the past ten years. In 1983 and 1984, price appears to have settled near the long-run equilibrium cost of mining coal, and few significant technological breakthroughs are likely to be implemented soon that would disturb this equilibrium. For northern and southern Appalachia alike, the fractions of costs accounted for by capital, labor, and materials are as follows:

| | Percentage of overall cost | | |
	Capital	Labor	Materials
Surface mining[7]	52	21	26
Underground mining	24	50	26

These estimates illustrate clearly that surface mining is far more capital-intensive than underground mining. Increases in wages that are not offset by equal increases in productivity have a far greater cost-push inflationary effect in underground operations. That is one reason why underground operators are highly sensitive to union demands, particularly during slack times, lest their product be priced out of the market.

A summary incorporating the most widely used projections of domestic coal consumption and exports for the remainder of the decade is given in table 75. The data span the years 1984-1990 and include five independent forecasts. Domestic consumption is divided into the three principal consumption categories: utility demand, industrial and retail demand, and metallurgical demand.

The divergence among the various projections is largely the result of variances in assumptions about future world oil prices, the level of economic activity, and selected socioeconomic and demographic factors. But there are also other differences in assumptions among the models, such as those concerning the cyclical behavior of the economy and the annual percentage change in real disposable personal income.

The consensus of the projections is that coal consumption by the electric power industry will grow at an average annual rate of 2.5% or less. This growth rate is considerably less than the projected rates of growth in the other markets. The rate of growth in electric power generation, and with it growth in the demand for coal, has changed dramatically in the past ten years, and the effects of these changes linger on. Most important is the continuing focus on energy conservation which started in the middle to late 1970s and will continue through the 1980s. At least one-half of the decline in the historic growth rate is attributable to conservation. Approximately one-fourth of the decline in the rate came as the result of the weakened state of the economy in recent years, and the remainder of the decline is explained by structural changes in the use of energy. These include the effects of the substitution on non-energy factor inputs—materials and labor—for energy inputs. In short, the surging price of electric power set into motion a search for less costly alternatives, including technological developments requiring fewer energy inputs. This trend is continuing, as is the growth of the service sector of the economy that in itself is less energy intensive than the basic industrial sectors. Structural changes in the economy are severing the link between economic growth and the growth of electric power. The former now proceeds with less input from the latter.

Some forecasting services, notably ICF Inc., project what may be an overly optimistic annual growth rate of 5.7% in the industrial consumption of coal; Chase Econometrics is much less optimistic, with a 2% projection. Overall, the demand for coal by each of the industrial categories depends upon growth of production in the sector and the electric utilities' conversion from oil- and gas-fired furnaces to coal-fired ones. Neither of these sources of additional future demand is likely to materialize in the 1980s. The costs of investment capital are predicted to remain high, and few forecasters expect a surge in industrial production in the next few years. Consequently the predicted rates of growth in demand for electric power appear realistic.

Projections of the demand for coking coal (the metallurgical market) for the remainder of the decade also appear somewhat optimistic, in particular when one considers the problems faced by the U.S. steel industry. Not only do domestic producers of steel face intense competition from imported steel, but they also face the threat that lighter and more resilient aluminum and plastic components will be substituted for steel in many products. Consequently, the average annual growth rate of 5.6% in the demand for coking coal appears exceedingly high. If the demand for metallurgical coal depends essentially on the demand for steel and only secondarily on the price of coal, projections for the use of this coal should not be optimistic. It is quite likely that an average annual output of approximately 40 million short tons will be the maximum amount that consumers of coking coal will buy for the remainder of the decade. Even this forecast must be tempered by the recognition that the use of coking coal per ton of steel has fallen gradually but consistently in recent years because of technological changes.

The last demand category included in the consumption forecasts of table 75 is exports. This category shows larger variations in estimates than any other demand category. Forecasts range from an average annual decline of nearly 1% to an increase of as much as 5.7%. The export market is clearly the most difficult to predict, not only because exports serve both the steam coal and metallurgical markets, but also because a host of unpredictable economic, political, and environmental factors can influence future coal flows dramatically.

The export boom of recent years, much to the chagrin of a disappointed Appalachian coal industry, was short-lived. The surge in steam coal exports was attributable to the widely publicized political turmoil in Poland and labor unrest in Australia. Faced with interrupted coal supplies and unmet contractual obligations, coal consumers in Europe and Japan turned to the only alternative supply source available, the U.S. spot market. That this situation was temporary was clearly in evidence from the outset because foreign buyers were unwilling to commit themselves to long-term contracts.

In Poland, coal production has been restored to its earlier levels, and coal is once again offered in the international market in large quantities. Against weak demand, prices have fallen. Moreover, Poland's convertible currency needs will be so acute for the remainder of the decade that coal, the nation's principal currency earner, will continue to be offered at discount prices. Price competition with an economy governed by a central-planning system is unrealistic because the selling price is not necessarily tied to the costs of production in such an economy. Poland can continue to sell coal below cost ad infinitum. Not a market force but an administrative decision sets the price, and that decision depends entirely on the central-planning authority's ranking of priorities. Exacerbating the situation for competing exporters is the fact that, unlike its coal-mining sector, Poland's industrial sector has been slow to recover from the dislocations and protests of the early 1980s. Hence, the domestic demand for coal is weak, putting further pressure on the coal sector to sell its output abroad at whatever prices it takes to find buyers.

Australia's labor problems are settled now, and the coal sector has once again resumed its traditional role of a competitive supplier in world markets. Consequently, in the middle 1980s, the world market for coal has become a buyer's market. United States exporters, whose coal is considerably more expensive delivered than that of its competitors, have seen their exports shrink markedly. Exacerbating this situation is, of course, the strength of the U.S. dollar abroad. To foreign buyers, the higher price of the dollar means that they have to give up significantly more of their own currency and resources for each dollar they buy to pay for U.S. coal. The terms of trade are decidedly unfavorable for foreign buyers of U.S. coal. The parity of European currencies with the currencies of other coal-exporting nations has in most cases not changed. Consequently, buying coal from sources other than the United States is an economically rational action at this time.

The future international value of the dollar is impossible to predict. However, because of the sizable U.S. budget deficits that now appear structurally entrenched, U.S. interest rates may remain high by historic standards for the remainder of the decade. This would, of course, keep the dollar at its present lofty levels, not an attractive prospect for coal exporters.

The preceding arguments in this chapter and the data in table 75 suggest that the markets for coal, and particularly for Appalachian coal, are influenced by many more varied, unpredictable, and complicated factors than are markets for most other products. No single prediction seems unquestionably reliable, and no scenario includes the impact of sudden and unexpected events. Few models, for instance, adequately consider the impact on coal demand of sharply declining prices of oil, although such prices have for some time headed downward uniformly and consistently. Analytical models provide reference points, at best. Sound conclusions about the future require that model forecasts be tempered by judgment, because many conditions and situations cannot be incorporated into equations and expressed quantitatively. Two respected researchers, recognizing that in recent years "seldom have so many knowledgeable observers been wrong so often," have suggested some "not unlikely" interactions between economics and energy.[8] Some of the views they present show a smoothly changing world during the 1980s, and others assume sudden shocks influencing prices. If one eliminates the extreme views of what might happen in the 1980s, what remains is a set of not unlikely combinations for the future. This chapter has included some likely and not necessarily optimistic views of the future. Although most models predict a relatively attractive outlook for U.S. and Appalachian coal, the future may be less optimistic than many have predicted. The crushing recession in the industrial world, along with continuing grievous levels of unemployment and (in some countries) unacceptable rates of inflation, has stifled the demand for coal. Recovery closely hinges upon a resurgence in worldwide economic activity, on favorable governmental regulation, and on social and economic stability.

APPENDIX

Table 1. World Coal Resources and Reserves
(millions of metric tons coal equivalents)

Country	Technically and Economically Recoverable Reserves[a]	Percent of Total	Geological Resources	Percent of Total
United States	166,950	25.2	2,570,398	23.9
U.S.S.R.	109,900	16.6	4,860,000	45.2
China	98,883	14.9	1,438,045	13.4
Poland	59,600	9.0	139,750	1.3
United Kingdom	45,000	6.8	190,000	1.8
South Africa	43,000	6.5	72,000	0.7
West Germany	34,419	5.2	246,800	2.3
Australia	32,800	4.9	600,000	5.5
India	12,427	1.9	81,019	0.8
Canada	4,242	0.6	323,036	3.0
Other countries	55,711	8.4	229,164	2.1
Total	662,932	100.0	10,750,212	100.0

Source: Derived from C.L. Wilson, *Coal: Bridge to the Future* (Cambridge, Mass.: Ballinger, 1980), p. 161.

[a]See note 2, chap. 1, for definition.

Table 2. Technically and Economically Recoverable
Reserves, by Type of Coal (millions of metric tons coal equivalent)

Country	Bituminous and Anthracite	Percent of Total	Subbituminous and Lignite	Percent of Total	Total	Percent of Total
United States	113	22.9	64	44.4	177	27.8
China	99	20.1	—		99	15.5
U.S.S.R.	83	16.8	27	18.7	110	17.3
United Kingdom	45	9.1	—		45	7.1
India	33	6.7	—		33	5.2
South Africa	27	5.5	—		27	4.2
Germany	24	4.9	11	7.6	35	5.5
Poland	20	4.1	1	0.7	21	3.3
Australia	18	3.7	9	6.3	27	4.2
Canada	9	1.8	1	0.7	10	1.6
Others	22	4.4	31	21.6	53	8.3
Total	493	100.0	144	100.0	637	100.0

Source: International Energy Agency, *Steam Coal* (Paris: Organization for Economic Cooperation and Development, 1978).

[a]Reserve estimates for United Kingdom are judged to be economical to extract only at some future date.

Table 3. Major Countries' Coal Production
(millions of short tons)

Country	1977	1978	1979	1980	1981	1982	1983	1984
United States	697	670	781	830	824	838	782	896
U.S.S.R.	796	798	792	790	776	791	789	787
China	606	681	698	684	683	734	788	850
East Germany	280	279	282	285	294	294	309	310
Poland	250	258	264	254	219	250	258	267
West Germany	229	228	239	239	241	247	236	233
Australia	111	114	119	116	130	140	146	158
South Africa	94	100	114	127	144	154	161	161
India	115	116	118	125	142	146	158	167
United Kingdom	135	136	135	141	138	134	127	56
Czechoslovakia	134	136	137	136	137	138	140	145

Source: U.S. Department of Energy, Energy Information Administration, *Weekly Coal Production* (Washington, D.C.: Government Printing Office, 1982). *Annual Energy Review, 1983*, Washington, D.C. 1984; *International Energy Annual, 1984*, Washington, D.C., October 1985.

Note: Includes anthracite, subanthracite, bituminous, subbituminous, lignite, and brown coal.

Table 4. Energy Consumption, Population, and Gross National Product
of the United States, 1950-1984

Year	Gross Energy Consumption (in quads) (1)	Population (in millions) (2)	GNP (in billions of 1972 dollars) (3)	GNP per Capita (in 1972 dollars) (4)	Btu per Dollar of GNP (in 1972 dollars) (5)	Millions of Btu per Capita (6)
1950	34,000	152.3	533	3,503	63.7	223.2
1955	39,700	165.9	654	3,947	60.6	239.3
1960	44,600	180.7	736	4,077	60.5	246.8
1965	53,500	194.3	925	4,765	57.6	274.3
1970	66,900	204.9	1,075	5,248	62.2	326.5
1973	74,212	210.4	1,254	5,870	59.2	352.7
1974	72,479	211.9	1,246	5,747	58.2	342.0
1975	70,485	213.6	1,232	5,628	57.2	330.0
1976	74,297	215.1	1,298	5,918	57.2	345.4
1977	76,215	217.8	1,370	6,152	55.6	349.4
1978	78,039	219.4	1,439	6,377	54.2	355.9
1979	78,845	221.1	1,479	6,707	53.3	356.6
1980	75,900	222.8	1,475	6,646	51.5	340.7
1981	73,940	227.0	1,514	6,652	48.8	325.7
1982	70,822	232.1	1,485	6,398	47.7	305.1
1983	70,573	234.2	1,535	6,554	46.6	301.3
1984	73,730	236.6	1,639[a]	6,927	45.0	311.6

Sources: Phillip G. LeBel, *Energy Economics and Technology* (Baltimore: Johns Hopkins Univ. Press, 1982), pp. 58-59 for 1950-1970; U.S. Department of Energy, Energy Information Administration, *Monthly Energy Review* (Washington, D.C.: Government Printing Office, 1982) for 1973-1983, and March 1985; U.S. Department of Commerce, Bureau of Census, *Statistical Abstract of the United States* (Washington, D.C.: Government Printing Office, 1985); U.S. Department of Commerce, *Survey of Current Business*, January 1985.

[a]Preliminary figure.

Table 5. Delivered Cost of Fossil Fuels
to Electric Utility Plants, 1973-1984
(cents per million Btu)

Year	Coal	Percent Change	Heavy Oil	Percent Change	Natural Gas	Percent Change	All Fossil Fuels
1973	40.5	—	78.5	—	33.8	—	47.6
1974	70.9	+ 75.1	189.0	+ 140.8	48.2	+ 42.6	91.4
1975	81.4	+ 14.8	200.5	+ 6.1	75.2	+ 56.0	104.4
1976	84.8	+ 4.2	195.2	− 2.6	103.4	+ 37.5	111.9
1977	94.7	+ 11.7	219.8	+ 12.6	129.1	+ 24.9	129.7
1978	111.6	+ 17.8	212.5	− 3.3	142.2	+ 10.1	141.1
1979	122.4	+ 9.7	298.8	+ 40.6	174.9	+ 23.0	163.9
1980	135.2	+ 10.5	426.7	+ 42.8	219.9	+ 25.7	192.8
1981	153.2	+ 13.3	533.4	+ 25.0	280.5	+ 27.6	225.6
1982	164.7	+ 7.5	483.2	− 9.4	337.6	+ 20.4	224.9
1983	165.6	+ 0.5	457.8	− 5.3	347.4	+ 2.9	220.6
1984	166.3	+ 0.4	481.0	+ 5.1	357.9	+ 3.0	219.2

Source: U.S. Department of Energy, Energy Information Administration, *Monthly Energy Review* (Washington, D.C.: Government Printing Office, 1985).

Note: All prices are yearly averages. Monthly fluctuation in fuel prices at times can be substantial.

Table 6. U.S. Gross Energy Consumption, by Type of Fuel,
1950-1984 (quadrillions of Btu)

Year	Coal	Petroleum	Natural Gas	Nuclear	Hydropower	Total
1950	12.91	13.50	6.20	0.0	1.40	34.01
1960	10.10	20.10	12.70	0.0	1.70	44.60
1970	12.70	29.50	21.80	0.02	2.70	66.72
1973	12.90	34.84	22.51	0.91	3.01	74.20
1974	12.60	33.46	21.73	1.27	3.31	72.48
1975	12.60	32.73	19.95	1.90	3.22	70.48
1976	13.52	35.18	20.35	2.11	3.07	74.30
1977	13.85	37.12	19.93	2.70	2.52	76.22
1978	13.71	37.97	20.00	3.02	3.14	78.04
1979	14.98	37.12	20.67	2.78	3.14	78.84
1980	15.37	34.20	20.39	2.74	3.12	75.90
1981	15.86	31.93	19.93	3.01	3.10	73.94
1982	15.24	30.23	18.51	3.12	3.59	70.82
1983	15.88	29.98	17.50	3.24	3.86	70.57
1984	17.20	31.00	18.03	3.55	3.78	73.73

Source: Same as table 4.

Table 7. Consumption of U.S. Coal by Sector, 1970-1984
(thousands of short tons)

Year	Electric Utilities	Percent of Total	Coke Plants	Other Industrial Including Transportation	Residential and Commercial	Total
1970	318,921	61.9	96,009	88,617	12,072	515,619
1973	389,212	69.2	94,101	68,617	11,117	562,584
1974	391,811	70.2	90,191	64,983	11,417	558,402
1975	405,962	72.2	83,598	63,670	9,410	562,641
1976	448,371	74.3	84,704	61,799	8,916	603,790
1977	477,126	76.3	77,739	61,472	8,954	625,291
1978	481,125	77.0	71,394	63,085	9,511	625,225
1979	527,051	77.4	77,368	67,717	8,388	680,524
1980	569,274	81.0	66,660	60,347	6,452	702,733
1981	596,797	81.5	61,014	67,395	7,422	732,627
1982	593,666	84.0	40,908	64,097	8,240	706,911
1983	625,569	85.1	37,025	64,361	8,470	735,425
1984a	664,399	84.0	44,022	73,745	9,130	791,296

Source: U.S. Department of Energy, Energy Information Administration, *Monthly Energy Review* (Washington, D.C.: Government Printing Office, 1985); U.S. Department of Energy, Energy Information Administration, *1984 Annual Energy Review* (Washington, D.C.: Government Printing Office, 1985).

aPreliminary figures.

Table 8. Demonstrated Reserve Base of Coal in the United States on January 1, 1983
(millions of short tons)

Area	Method of Mining	Anthracite	Bituminous	Subbituminous	Lignite	Total[a]
States East of the	Underground	7,085.6	180,402.7	—	—	187,488.3
Mississippi River	Surface	113.5	39,382.7	—	1,083.0	40,579.2
	Total	7,199.1	219,785.4	—	1,083.0	228,067.5
States West of the	Underground	116.4	26,696.8	118,813.4	—	145,626.6
Mississippi River	Surface	15.8	9,098.1	62,670.6	44,036.6	115,820.9
	Total	131.9	35,794.9	181,484.0	44,036.6	261,447.4
United States	Underground	7,202.0	207,099.4	118,813.4	—	333,114.9
Overall	Surface	129.0	48,480.9	62,670.6	45,119.6	156,400.1
	Total	7,331.0	255,580.3	181,484.0	45,119.6	489,514.9

Source: U.S. Department of Energy, Energy Information Administration, *Coal Production 1983* (Washington, D.C.: Government Printing Office, October 1984).

[a]Data may not equal totals because of rounding.

Table 9. Geographic Distribution of Demonstrated Reserve Base
by Type of Mining, January 1, 1983
(millions of short tons)

States[a]	Underground	Surface	Total
Alabama	1,717.4	3,443.2	5,160.6
Alaska	5,423.0	728.6	6,151.6
Arizona	101.6	264.9	366.5
Arkansas	272.5	145.4	417.9
Colorado	12,258.8	4,936.2	17,195.0
Georgia	1.9	1.7	3.6
Idaho	4.4	—	4.4
Illinois	63,426.6	15,625.1	79,051.7
Indiana	8,931.6	1,552.5	10,484.1
Iowa	1,733.6	461.6	2,195.2
Kansas	—	989.1	989.1
Kentucky			
Eastern	17,186.0	2,130.4	19,316.4
Western	16,885.8	3,999.3	20,885.1
Maryland	702.3	101.5	803.8
Michigan	123.1	4.6	127.7
Missouri	1,479.1	4,570.4	6,049.5
Montana	70,958.7	49,355.1	120,313.8
New Mexico	2,128.0	2,549.1	4,677.1
North Carolina	10.7	—	10.7
North Dakota	—	9,886.2	9,886.2
Ohio	12,995.1	5,895.2	18,890.3
Oklahoma	1,238.3	379.5	1,617.8
Oregon	14.5	2.9	17.4
Pennsylvania	28,440.2	1,593.7	30,033.9
South Dakota	—	366.1	366.1
Tennessee	627.7	312.6	940.3
Texas	—	13,812.5	13,812.5
Utah	6,121.7	267.9	6,389.6
Virginia	2,435.8	814.4	3,250.2
Washington	1,332.3	131.5	1,463.8
West Virginia	34,004.2	5,105.1	39,109.3
Wyoming	42,560.1	27,004.1	69,564.2
Total	333,115.0	156,400.4	489,515.4

Source: Same as table 8.

Note: Table values include those parts of measured and indicated resource categories as defined by the Energy Information Administration and represent 100% of the coal in place.

[a]Excludes coal-bearing states in which either the resources are not currently economically recoverable or the publicly available resource data do not provide the detail required for DRB delineation.

Table 10. Appalachian Coal Basin Reserves, January 1, 1983
(millions of short tons)

State or Area	Underground	Surface	Total
Alabama	1,717.4	3,443.2	5,160.6
Eastern Kentucky	17,186.0	2,130.4	19,316.4
North Carolina	10.7	—	10.7
Ohio	12,995.1	5,895.2	18,890.3
Pennsylvania	28,440.2	1,593.7	30,033.9
Tennessee	627.7	312.6	940.3
Virginia	2,435.8	814.4	3,250.2
West Virginia	34,004.2	5,105.1	39,109.3
Total	97,417.1	19,294.6	116,711.7
Total as a percentage of U.S. reserves	29.2	12.3	23.8

Source: Derived from table 9.

Table 11. Coal Characteristics of Principal Producing
States in Appalachia

State or Area	Rank	Percentage		Btu/lb.
		Ash	Sulfur	
Eastern Kentucky	High-volatile bituminous, some good coking coal	3.5– 11.0	0.6–3.3	12,300–14,200
West Virginia	High-volatile bituminous, much good coking coal	2.0–28.0	0.4–14.0	10,200–15,600
Pennsylvania	High-volatile bituminous and anthracite, much good coking coal	4.1–15.1	0.5–4.2	12,580–14,490
Ohio	High-volatile bituminous	6.6–11.5	1.5–5.0	11,006-12,919

Source: Committee on Science and Technology, *Energy Facts.*

Table 12. U.S. Coal Production
(millions of short tons)

Year	Tons
1960	413
1970	603
1975	655
1979	781
1980	830
1981	824
1982	838
1983	782
1984	890 [a]

Source: U.S. Department of Energy,
Energy Information Administration,
Weekly Coal Production, January 29,
1983 (Washington, D.C.: Government
Printing Office, 1983) and Department
of Energy, Energy Information
Administration, *Monthly Energy
Review* (Washington, D.C.:
Government Printing Office, 1984).
Quarterly Coal Report (April 1985).

Note: Production figures include
bituminous coal, lignite, and
anthracite.

[a]Preliminary estimate.

Table 13. Bituminous Coal Production by State and Region, 1960-1984
(millions of short tons)

State and Region	1960	1970	1975	1979	1980	1981	1982	1983	1984[a]
Appalachian Coal Basin									
Alabama	12.6	20.6	22.6	24.2	26.4	24.5	26.6	23.8	27.1
Eastern Kentucky	36.3	72.5	82.3	104.1	109.2	117.9	111.2	95.6	124.8
Maryland	0.8	1.6	2.6	2.6	3.8	4.4	3.8	3.2	4.2
Ohio	34.0	55.4	46.8	43.6	39.4	37.4	36.5	33.8	39.0
Pennsylvania[b]	65.5	80.5	84.1	93.3	87.1	78.1	74.8	65.7	71.1
Tennessee	5.0	8.2	8.2	8.7	9.9	10.6	7.5	6.6	7.7
Virginia	29.0	35.0	35.5	36.8	41.0	42.0	39.8	35.0	35.3
West Virginia	119.5	144.1	109.3	112.1	121.6	112.8	128.5	115.0	129.6
Total	302.7	417.9	391.4	425.4	438.4	427.7	428.7	378.7	438.8
Eastern Interior Coal Basin									
Western Kentucky	30.6	52.8	56.4	42.5	40.9	39.7	39.0	35.6	40.7
Illinois	46.0	65.1	59.6	59.6	62.5	51.9	60.3	56.8	65.3
Indiana	15.1	22.3	25.1	27.5	30.8	29.3	31.8	31.8	36.8
Total	91.7	140.2	141.1	129.6	134.2	120.9	133.1	127.2	142.8
Western coal region									
Arizona		0.1	7.0	11.4	10.9	11.6	12.4	11.4	11.7
Colorado	3.6	6.0	8.2	18.5	18.8	19.9	18.3	16.7	16.8
Montana	0.3	3.5	22.1	32.7	29.9	33.6	27.7	28.9	32.8
New Mexico	0.3	7.4	8.8	15.6	18.4	18.7	19.9	20.4	24.6
Utah	4.9	4.7	7.0	12.0	13.2	13.8	17.0	11.8	13.1
Wyoming	2.0	7.2	23.8	71.5	94.9	103.0	108.4	112.2	128.4
Total	11.1	28.9	76.9	161.7	186.1	200.6	203.7	201.4	227.4

Source: U.S. Department of Energy, Energy Information Administration, *Weekly Coal Production, November 6, 1982, January 21, 1984, January 25, 1985* (Washington, D.C.: Government Printing Office, 1982, 1984, 1985).

[a]Preliminary data.
[b]Exclusive of anthracite.

Table 14. Estimated Long-run Price Elasticities of Demand
for Oil, Natural Gas, and Coal, by Demand Sector

Demand Sector	Oil	Natural Gas	Coal
Electricity-generating sector			
Griffin (1977)	−1.00 to −4.00	−0.80 to −1.20	−0.50 to −0.80
Atkinson-Halverson (1976)	−1.50 to −1.60	−1.40	−0.40 to −1.20
Industrial sector			
Baughman-Zerhoot (1975)	−1.32 to −1.40	−0.81 to −1.51	−0.59 to −1.14
Pindyck (1979)	−0.20 to −1.20	—	−1.30 to −2.20
Residential and commercial sector			
Baughman-Joskow (1974)	−0.81	−0.62	——
Anderson (1972)	−1.58	−1.73	——
Joskow-Baughman (1976)	−1.00 to −1.10	−1.00 to −1.10	——
Hirst-Lin-Cope (1976)	−0.84 to −0.91	−0.84 to −0.91	——
Pindyck (1979)	−1.10 to −1.30	−1.30 to −2.10	——

Sources: M.L. Baughman and P.L. Joskow, "Energy Consumption and Fuel Choices by Residential and Commercial Users in the United States" (Cambridge, Mass.: MIT Energy Laboratory, July 1974); K.P. Anderson, "Residential Demand for Electricity: Econometric Estimates for California and the United States" (Santa Monica, Calif.: Rand Corp., 1972); M.L. Baughman and F.S. Zerhoot, "Energy Consumption and Fuel Choice by Industrial Consumers in the United States" (Cambridge, Mass.: MIT Energy Laboratory, March 1975); P.L. Joskow and M.L. Baughman, "The Future of the U.S.Nuclear Energy Industry," Bell Journal of Economics vol. 7, no. 1 (spring 1976); J.M. Griffin, "Interfuel Substitution Possibilities: A Translog Application to Pooled Data," International Economic Review vol. 18, no. 3 (October 1977); S.E. Atkinson and R. Halverson, "Interfuel Substitution in Steam Electric Power Generation," Journal of Political Economy vol. 84, no. 5 (October 1976); R.S. Pindyck, The Structure of World Energy Demand (Cambridge, Mass.: MIT Press, 1979); E. Hirst, W. Lin, and J. Cope, "An Engineering Economic Model of Residential Energy Use," Technical Report, TM 5470 (Oak Ridge, Tenn: Oak Ridge National Laboratory, July 1976); and P.G. LeBel, Energy Economics and Technology (Baltimore: Johns Hopkins Univ. Press, 1982).

Table 15. Demand Elasticities for U.S. Coal, 1952-1977

Consumer Group and Geographic Region	Own Price		Cross Price				Final Product			
			Oil		Gas		Electricity		Pig Iron	
	Long Run	Short Run	Long Run	Short Run	Long Run	Short Run	Long Run	Short Run	Long Run	Short Run
Utilities										
Eastern U.S.	−0.56[a]	−0.42[a]	0.09	0.07	−0.03	−0.02[a]	0.88[a]	0.66[a]	—	—
Midwestern U.S.	−0.91[a]	−0.51[a]	.13	0.07	0.75[a]	0.42[a]	1.54[a]	0.86[a]	—	—
Industry-railroad-retail										
Eastern U.S.	−1.03	−0.41	0.03	0.01	0.05	0.02	—	—	—	—
Midwestern U.S.	−1.21[a]	−0.63[a]	0.02	0.01	0.35	0.18	—	—	—	—
Coke manufacturers										
Eastern U.S.	0.11	0.09	—	—	—	—	—	—	0.95[a]	0.78[a]
Midwestern U.S.	0.01	0.01	—	—	—	—	—	—	0.88[a]	0.68[a]
Western U.S.	0.18	0.16	—	—	—	—	—	—	0.98[a]	0.85[a]
Exports										
Japan	−1.81	−1.07	—	—	—	—	−3.08	1.82	3.39[a]	2.00[a]
Canada	0.78	0.26	—	—	—	—	−1.06	−0.35	1.67[a]	0.55[a]
Rest of the world	−0.18	−0.14	—	—	—	—	—	—	1.57[a]	1.21[a]

Source: M.M. Ali, C.E. Harvey, and J.F. Stewart, *Regional Demand and Supply Behavior by Sectors of the U.S. Coal Industry* (Lexington: Univ. of Kentucky, Institute for Mining and Minerals Research, 1981).

[a]Coefficient magnitude exceeds two standard errors.

Table 16. Price Elasticity of Supply of the Fossil Fuels

Fuel	Short-Run Price Elasticity	Long-run Price Elasticity
Petroleum	0.111 to 0.141[a]	0.34 to 0.67; 0.37 to 0.42[a]
Natural gas	0.042	0.27 to 0.30; 0.19 to 0.35[b]
Coal	0.142	0.21 to 0.32

Source: Phillip G. LeBel, *Energy Economics and Technology* (Baltimore: Johns Hopkins Univ. Press, 1982), p. 251.

[a]These estimates are based on North Sea discoveries; thus, conditions are not directly comparable to U.S. conditions.

[b]Data are for 1985, using 1975 constant dollars per 1,000 cubic feet within a range of $1.00 to $2.80.

Table 17. Largest U.S. Commercial Bituminous Coal Producers, 1983

Rank	Coal Firm	Ownership or Controlling Company	1983 Commercial Production	Percent of Total Commercial Production[a]	Cumulative Percent
1	Peabody Group	Peabody Holding Co.	53,430,000	6.85	6.85
2	Consolidation Group	Continental Oil Co.	42,200,000	5.41	12.26
3	Amax Group	Amax, Inc.	40,016,235	5.13	17.39
4	Texas Utilities	Texas Utilities	28,355,053	3.64	21.03
5	Exxon Coal USA	Exxon Coal USA	19,437,000	2.49	23.52
6	A.T. Massey Group	A.T. Massey	18,218,000	2.34	25.86
7	Anaconda Minerals	Anaconda Minerals	18,004,123	2.31	28.17
8	Utah International	Broken Hill Proprietary Co.	15,150,000	1.94	30.11
9	Kerr-McGee Coal Corp.	Kerr-McGee Corp.	14,900,000	1.91	32.02
10	NERCO	Pacific Power & Light Co.	14,802,000	1.90	33.92
11	North American	North American Coal Co.	14,358,241	1.84	35.76
12	Island Creek Group	Occidental Petroleum Corp.	14,123,508	1.81	37.57
13	Pittston Coal	Pittston Co.	13,472,630	1.73	39.30
14	American Electric Power	American Electric Power	13,262,275	1.70	41.00
15	Westmoreland Group	Westmoreland Coal Co.	12,591,985	1.61	42.61
	Total		332,321,050	42.61	

Source: Compiled from *Keystone Coal Industry Manual* (New York: McGraw-Hill, 1984).
[a]Total commercial production during 1983 was 780,000,000 tons.

Table 18. 1981 Estimated Coal Tonnage Held in Reserves by
Individual Firms
(billions of short tons)

Type of Ownership[a]	Numbers of Companies	Tons	Holdings in excess of 1 billion	All Holdings
			Percent of Total	
Oil	12	51.523	41.4	32.0
Railroads	3	26.100	21.0	16.2
Coal	5	10.400	8.4	6.4
Utilities	4	5.450	4.4	3.4
Steel	3	5.100	4.1	3.2
Other (metals, gas, defense)	10	25.752	20.7	16.0
Subtotal	37	125.325	100.0	77.2
Holdings over 1 billion tons each	393	35.921		22.3
Total	430	161.246		99.5

Source: Same as table 17.

Note: Data are from organizations owning, leasing, or controlling coal reserves in the United States and Canada. No differentiation is made between recoverable and in-place reserves.

[a]Firms included have holdings in excess of 1 billion tons of annual production.

Table 19. Percentage Share of Bituminous Coal Production by
Region, 1960-1984

Producing Region	1960	1970	1975	1979	1980	1981	1982	1983	1984
Appalachian basin	75	71	64	59	58	57	56	54	54
Eastern basin	23	24	23	18	18	16	17	18	18
Western region	2	5	13	23	24	27	27	28	28
Total	100	100	100	100	100	100	100	100	100

Source: Calculated from table 13.

Table 20. Distribution of Appalachian Coal by End User, 1970-1984

Quantity and User	1970	1973	1975	1979	1980	1981	1982	1983	1984
Millions of short tons used									
Domestic									
Electric utilities	193,092	195,664	218,285	238,741	245,498	235,879	241,057	235,276	271,213
Coke	92,252	87,691	81,859	68,877	58,359	51,165	35,280	33,831	39,325
Other uses (retail, industrial, commercial, transportation)	73,979	50,891	39,901	40,018	39,554	39,532	36,590	40,226	50,044
Exports									
Canada	—	—	—	17,810	15,845	14,852	16,172	15,805	19,312
Elsewhere	51,316	35,570	48,182	47,131	72,828	83,397	81,362	56,470	58,908
Total	410,641	369,825	388,227	412,587	432,084	424,827	410,462	381,608	438,802
Percentage of total coal used									
Electric utilities	47	52	56	58	57	56	58	62	62
Coke	22	24	21	17	13	12	9	9	9
Other uses (retail, industrial, commercial, transportation)	18	14	10	10	9	9	9	10	11
Exports									
Canada	—	—	—	4	4	3	4	4	4
Elsewhere	13	10	13	11	17	20	20	15	13
Total	100	100	100	100	100	100	100	100	99

Source: U.S. Department of Interior, Bureau of Mines, *Mineral Industry Surveys, 1970, 1973, and 1975* and U.S. Department of Energy, Energy Information Administration, *Coal Distribution, 1979-1984.* Both published in Washington, D.C., by Government Printing Office (1970, 1973, and 1975 for the former and 1979- 1983 for the latter.)

Note: For 1970, 1973, and 1975, shipments to Canada and Mexico are included in the U.S. shipments. For 1979-1984, Canadian shipments are reported separately and Mexican shipments are included in exports elsewhere. Columns may not add to total because of rounding.

Table 21. Average U.S. Price of Electricity Sold to End Users, 1960-1984 (cents per kilowatt hour)

Year	Residential		Commercial		Industrial		Other		Weighted Average	
	Current	Constant	Current	Constant	Current	Constant	Current	Constant	Current	Constant
1960	2.62	3.81	2.42	3.52	1.06	1.54	1.91	2.78	1.82	2.65
1965	2.39	3.21	2.18	2.93	1.00	1.34	1.82	2.45	1.70	2.29
1970	2.22	2.43	2.08	2.27	1.02	1.12	1.80	1.97	1.67	1.83
1971	2.32	2.42	2.20	2.29	1.10	1.15	1.91	1.99	1.77	1.84
1972	2.42	2.42	2.29	2.29	1.16	1.16	1.98	1.98	1.86	1.86
1973	2.54	2.40	2.41	2.28	1.25	1.18	2.10	1.99	1.96	1.85
1974	3.10	2.69	3.04	2.64	1.69	1.47	2.75	2.39	2.49	2.16
1975	3.51	2.79	3.45	2.74	2.07	1.65	3.08	2.45	2.92	2.32
1976	3.73	2.82	3.69	2.79	2.21	1.67	3.27	2.47	3.09	2.33
1977	4.05	2.89	4.09	2.92	2.50	1.79	3.51	2.51	3.42	2.44
1978	4.31	2.87	4.36	2.90	2.79	1.85	3.62	2.41	3.69	2.45
1979	4.64	2.84	4.68	2.86	3.05	1.87	3.96	2.42	3.99	2.44
1980	5.36	3.00	5.48	3.07	3.69	2.07	4.76	2.66	4.73	2.65
1981	6.20	3.17	6.29	3.22	4.29	2.19	5.28	2.70	5.46	2.79
1982	6.86	3.31	6.86	3.31	4.95	2.39	5.92	2.86	6.13	2.96
1983	7.18	3.33	7.01	3.25	4.97	2.30	6.36	2.95	6.29	2.92
1984	7.56	3.43	7.32	3.32	5.03	2.28	6.77	3.07	6.52	2.96

Source: Energy Information Administration, *1982 Annual Energy Review* (Washington, D.C.: Government Printing Office, 1983); *Monthly Energy Review, January, 1985* (Washington, D.C.: Government Printing Office, 1985).

Note: Constant dollars are 1972 dollars using the GNP implicit price deflator.

Table 22. U.S. Production of Electricity by Type of Energy Source,
1960-1984 (billions of kilowatt hours)

Year	Coal	Petroleum	Natural Gas	Nuclear	Hydro-power	Geothermal and Other	Total
1960	403	46	158	1	146	0	754
1965	571	65	222	4	194	0	1,055
1970	704	184	373	22	248	1	1,532
1971	713	220	374	38	266	1	1,613
1972	771	274	376	54	273	2	1,750
1973	848	314	341	83	272	2	1,861
1974	828	301	320	114	301	3	1,867
1975	853	289	300	173	300	3	1,918
1976	944	320	295	191	284	4	2,038
1977	985	358	306	251	220	4	2,124
1978	976	365	305	276	280	3	2,206
1979	1,075	304	329	255	280	4	2,247
1980	1,162	246	346	251	276	6	2,286
1981	1,203	206	346	273	261	6	2,295
1982	1,193	147	305	283	309	5	2,242
1983	1,259	144	274	294	332	6	2,310
1984	1,342	120	297	328	321	9	2,417

Source: Same as table 21.

Table 23. Demand for Appalachian Bituminous Coal by Consumer
Category, 1984 (thousands of short tons)

Category	District 8[a] Tons	Percent of Total	Remainder of Appalachia[b] Tons	Percent of Total
Electric utilities	125,561	60	145,652	63
Coke plants	18,709	9	20,616	9
Other uses[c]	28,134	14	21,908	9
Exports[d]	35,706	17	42,514	19
Total	208,110	100	230,690	100

Source: U.S. Department of Energy, Energy Information Administration, *Coal Distribution, 1984* (Washington, D.C.: Government Printing Office, 1985).

[a]District 8 includes Eastern Kentucky, North Carolina, Tennessee, and portions of West Virginia and Virginia. Data include demand for anthracite coal.
[b]Includes Pennsylvania, Ohio, Alabama, portions of West Virginia, and Virginia.
[c]Other uses include industrial, residential, commercial, and transportation.
[d]Export estimates include coal going to Canada and Mexico.

Table 24. Consumption of U.S. Steam Coal in Electric Plants, by Region
(1972, 1977, 1982)

Region	Billions of Kilowatt Hours Generated			Percentage of Btu Provided by[a]								
				Coal			Oil			Gas		
	1982	1977	1972	1982	1977	1972	1982	1977	1972	1982	1977	1972
New England	45.7	45.3	52.1	24.4	5	6	72.3	94	93	3.1	1	1
Middle Atlantic	179.2	182.4	181.8	65.3	57	52	24.2	42	44	10.3	1	4
East North Central	306.4	319.1	290.4	97.7	95	91	1.8	4	4	0.3	1	5
West North Central	132.0	115.2	88.8	95.7	85	62	0.4	3	2	3.8	12	36
South Atlantic	313.1	301.6	266.6	81.1	70	62	13.4	26	29	5.4	4	9
East South Central	147.0	158.0	133.0	94.6	92	89	0.3	6	2	5.0	2	9
West South Central	293.5	253.8	198.6	38.0	9	1	0.7	9	2	61.2	82	97
Mountain	133.3	105.4	59.4	93.3	82	58	0.4	5	4	6.1	13	38
Pacific	63.1	130.3	88.1	9.8	12	0	14.8	61	30	75.3	27	70
U.S. total	1,613.3	1,612.1	1,358.8	66.7	62	54	14.3	20	19	18.9	18	27

Source: National Coal Association, *Steam Electric Plant Factors, 1972, 1977, and 1983* (Washington, D.C.: National Coal Association, 1973, 1978, 1983).

[a]Percentage may not add to 100 because of independent rounding.

Table 25. Percentage of Btu Derived from Gas, Coal, and Oil by Electric Utilities on Eastern Seaboard (1966, 1972, 1977, 1982)

State or City	1966			1972			1977			1982		
	Coal	Gas	Oil	Coal	Gas	Oil	Coal	Gas	Oil	Coal	Gas	Oil
Maine	0	0	100	0	0	100	0	0	100	0	0	100
New Hampshire	45	0	55	68	0	32	59	0	41	62	0	38
Massachusetts	45	5	50	1	0	97	0	1	99	30	5	65
Rhode Island	62	1	37	0	1	99	0	0	100	0	14	86
Connecticut	84	0	16	1	0	99	0	0	100	0	0	100
New York City	35	19	46	1	10	89						
New York State	99	1	0	57	4	39	23	1	76	25	22	53
New Jersey	61	7	32	10	7	83	30	2	68	43	30	27
Philadelphia	76	0	24	28	1	71	—	—	—	—	—	—
Pennsylvania	100	0	0	99	0	1	87	0	13	94	0	6
Delaware	87	12	1	45	4	51	30	2	68	60	7	33
Maryland	99	0	1	46	3	51	54	0	46	79	0	21
District of Columbia	98	0	2	14	0	86	0	0	100	0	0	100
Virginia	99	1	0	44	1	55	44	0	56	89	1	10
North Carolina	99	1	0	95	1	4	100	0	0	100	0	0
South Carolina	79	20	1	84	11	5	82	1	17	99	0	1
Georgia	100	0	0	80	12	8	90	1	8	100	0	0
Florida	20	27	53	18	24	58	23	19	58	30	23	47

Source: Calculated from National Coal Association, *Steam Electric Plant Factors, 1967-1983* (Washington, D.C.: National Coal Association, 1968 through 1984), table 2.

Table 26. Percentages of Total Regional Electricity Generation
by Nuclear Power

Region and State	1975	1982	January 1985
New England	30.9	24.6	35.2
Connecticut	39.7	37.7	54.7
Maine	74.0	55.1	64.3
Massachusetts	12.5	3.4	11.2
New Hampshire	0.0	0.0	0.0
Rhode Island	0.0	0.0	0.0
Vermont	97.5	80.6	77.5
Middle Atlantic	15.3	15.9	22.3
New Jersey	13.4	30.8	39.7
New York	16.6	17.7	25.5
Pennsylvania	14.9	11.1	15.5
East North Central	11.2	13.3	17.1
Illinois	24.2	21.8	39.6
Indiana	0.0	0.0	0.0
Michigan	10.5	19.7	19.6
Ohio	0.0	4.7	1.2
Wisconsin	32.4	29.3	32.5
South Atlantic	14.2	18.5	26.5
Delaware	0.0	0.0	0.0
District of Columbia	0.0	0.0	0.0
Florida	10.8	17.6	30.5
Georgia	8.3	9.4	9.1
Maryland	17.9	37.4	40.0
North Carolina	3.0	15.2	31.2
South Carolina	56.9	31.1	55.7
Virginia	26.4	51.2	50.9
West Virginia	0.0	0.0	0.0
East South Central	1.9	10.0	16.7
Alabama	6.3	25.8	25.7
Kentucky	0.0	0.0	0.0
Mississippi	0.0	0.0	10.9
Tennessee	0.0	4.5	22.8

Source: Data for 1975 were found in Edison Electric Institute (EEI), *Statistical Year Book, 1975* (New York, N.Y.: Edison Electric Inc., 1976), p. 23. Data for 1982 and 1983 are from Energy Information Administration (EIA), *Electric Power Monthly* (Washington, D.C.: Government Printing Office, 1983, 1985), p. 17.

Table 27. Coal Shipments from Several Appalachian States
to Steam-Electric Plants Producing at Least
25 Megawatts (thousands of tons), 1973-83

Year	Eastern Kentucky	West Virginia	Virginia	Ohio	Pennsylvania
1973	35,854.9	47,301.9	11,260.0	37,223.5	39,583.6
1974[a]	43,763.2	41,535.8	12,772.7	36,114.2	41,807.5
1975	47,315.0	43,923.7	13,869.0	39,606.5	42,003.3
1976	52,722.1	44,538.4	13,642.2	40,134.6	40,940.2
1977	61,577.6	44,166.1	13,745.6	41,654.9	44,368.2
1978[a]	62,803.1	38,301.8	12,140.4	34,561.3	42,036.7
1979	70,261.3	50,952.1	13,679.0	38,763.0	47,347.9
1980	73,938.0	53,115.2	13,769.1	34,323.9	49,964.7
1981[a]	76,184.3	50,872.9	12,110.4	31,453.3	46,024.9
1982	69,717.7	60,817.5	14,477.6	31,099.1	45,466.0
1983	61,597.9	65,560.2	14,034.9	30,078.7	44,790.6

Source: Federal Energy Regulatory Commission (formerly Federal Power
Commission), *Annual Summary of Cost and Quality of Electric Utility Plant Fuels,
1973-1977* (Washington, D.C.: Government Printing Office, 1973-1977) and U.S.
Department of Energy, Energy Information Administration, *Cost and Quality of
Fuels for Electric Utility Plants, 1978-1982* (Washington, D.C.: Government
Printing Office, 1978-1984).

[a]Strike year.

Table 28. Shipments of Coal (thousands of net tons) from Eastern Kentucky to Electric Utility Plants Producing 25 MW or More, 1983

Region and State	Tons Shipped	Percent of Total Eastern Kentucky Shipments to Utilities	Percent of Total Utility Shipments Received
Middle Atlantic			
New Jersey	31.1	0.0	1.2
New York	258.0	0.4	3.6
Pennsylvania	27.3	0.0	0.0
Total	316.4	0.5	
South Atlantic			
Florida	4,146.4	6.7	29.2
Georgia	9,834.9	16.0	37.9
North Carolina	7,737.2	12.6	39.0
South Carolina	6,210.2	10.0	81.9
Virginia	2,155.0	3.5	24.8
West Virginia	981.0	1.6	4.8
Total	31,064.7	50.5	
East North Central			
Illinois	1,048.6	1.7	3.2
Indiana	3.5	0.0	0.0
Michigan	8,405.8	13.7	36.0
Ohio	6,699.0	10.9	12.6
Wisconsin	31.0	0.0	0.2
Total	16,187.9	26.3	
East South Central			
Alabama	133.7	0.2	0.7
Kentucky	8,443.0	13.7	34.8
Mississippi	970.1	1.6	21.9
Tennessee	4,398.4	7.1	22.4
Total	13,945.2	22.7	
West North Central			
Minnesota	3.0	0.0	0.0
Iowa	9.1	0.0	0.1
Total	12.1	0.0	
Grand Total	61,526.3	100.0	

Source: Energy Information Administration, U.S. Department of Energy, *Cost and Quality of Fuels for Electric Utility Plant, 1983* (Washington, D.C.: Government Printing Office, 1984.)

Table 29. Rank-size Distribution of Electric Utility Market
for Eastern Kentucky Coal, 1983

State	Rank	Thousands of Net Tons	Percent of Total	Cumulative Percent
Georgia	1	9,834.9	16.0	16.0
Kentucky	2	8,443.0	13.7	29.7
Michigan	3	8,405.8	13.7	43.4
North Carolina	4	7,737.2	12.6	56.0
Ohio	5	6,699.0	10.9	66.9
South Carolina	6	6,210.2	10.0	76.9
Tennessee	7	4,398.4	7.1	84.0
Florida	8	4,146.4	6.7	90.7
Virginia	9	2,155.0	3.5	94.2
Illinois	10	1,048.6	1.7	95.9
West Virginia	11	981.0	1.6	97.5
Mississippi	12	970.1	1.6	99.1
New York	13	258.0	0.4	99.5
Alabama	14	133.7	0.2	99.7
New Jersey	15	31.1	0.0	99.7
Wisconsin	16	31.0	0.0	99.7
Pennsylvania	17	27.4	0.0	99.7
Iowa	18	9.1	0.0	99.7
Indiana	19	3.5	0.0	99.7
Minnesota	20	3.0	0.0	99.7

Source: Calculated from table 28.

Table 30. Domestic Shipments and Exports of District 8
Coal to Electric Utilities, 1970-1983 (millions of short tons)

Year	Total Coal Shipments	Shipments to Domestic Electric Utilities	Percent of Total	Exports as Percent of Total
1970	161.0	62.0	38.5	22.8
1971	150.4	69.8	46.4	17.0
1972	154.2	71.3	46.2	15.7
1973	147.4	65.7	44.6	16.0
1974	155.5	70.3	45.2	18.5
1975	159.5	79.6	49.9	18.6
1976	160.7	83.6	52.0	16.5
1977	159.0	90.2	56.7	13.1
1978	152.5	94.8	62.2	9.6
1979	173.8	103.2	59.4	15.4
1980	192.9	111.6	57.8	19.8
1981	202.6	113.4	56.0	22.3
1982	192.9	111.7	57.9	23.5
1983	179.7	106.1	59.0	19.7

Source: Kentucky Economic Information System, "Mineral Industry Surveys,"
derived from U.S. Department of the Interior, Bureau of Mines, "Energy Data
Report" (Washington, D.C.: U.S. Department of Energy, Energy Information
Administration).

Table 31. Appalachian Coal Shipments to Electric Utilities,
by State of Destination, 1982

State	Thousands of Tons	Percent of Total Coal Receipts
Alabama	15,962.2	91
Delaware	1,930.2	100
Florida	4,665.9	47
Georgia	12,618.7	60
Illinois	855.6	3
Indiana	273.8	1
Kentucky	10,497.5	41
Maryland	5,785.5	100
Massachusetts	3,089.0	100
Michigan	16,469.4	78
Minnesota	32.7	0
Mississippi	1,403.2	37
New Hampshire	862.0	100
New Jersey	2,878.6	100
New York	6,581.9	100
North Carolina	24,917.3	100
Ohio	42,408.5	94
Pennsylvania	39,048.4[a]	98
South Carolina	8,544.7	100
Tennessee	8,161.7	44
Vermont	39.6	100
Virginia	6,798.5	100
West Virginia	30,051.4	100
Wisconsin	995.6	7
Total	244,871.9	

Source: U.S. Department of Energy, Energy Information Agency, *Cost and Quality of Fuels for Electric Utility Plants* (Washington, D.C.: Government Printing Office, 1983).

Note: Discrepancy of total with total shown in table 21 is due to different reporting agencies.

[a]Bituminous coal shipments only. The remaining 2% is Pennsylvania anthracite coal.

Table 32. U.S. Exports of Bituminous Coal
(thousands of short tons)

Year	Metallurgical	Steam	Total
1973	42,607	10,263	52,870
1974	51,594	8,332	59,926
1975	51,597	14,072	65,669
1976	47,804	11,602	59,406
1977	41,891	11,796	53,687
1978	29,848	9,977	39,825
1979	50,698	14,085	64,782
1980	63,103	26,779	89,883
1981	65,234	45,010	110,243
1982	64,585	40,659	105,244
1983	49,964	26,905	76,869
1984	56,975	23,818	80,793

Source: U.S. Department of Energy, Energy Information Administration, *Weekly Coal Production, March 9, 1985* (Washington, D.C.: Government Printing Office, 1985), p. 13.

Note: Data exclude anthracite coal.

Table 33. U.S. Coal Exports to Each Continent, 1981-1984
(thousands of short tons)

Destination	1981	1982	1983	1984
Europe	57,012	51,211	33,062	32,810
Asia	31,959	29,957	22,254	21,600
North America	18,788	18,663	17,341	20,641
South America	3,419	3,956	4,130	5,407
Africa	1,337	1,457	965	1,023
Oceania and Australia	—	224	20	1
Total	112,515	105,468	77,772	81,482

Source: U.S. Department of Energy, Energy Information Administration, *Quarterly Coal Report, April 1983,* and *Weekly Coal Production, March 2, 1984, Quarterly Coal Report, April 1985* (Washington, D.C.: Government Printing Office, 1983, 1984, and 1985).

Table 34. Principal Foreign Markets of U.S. Coal, 1984

Destination	Thousands of Tons	Percent of Total Exports
Canada	20,444	25
Japan	16,325	20
Italy	7,624	9
Netherlands	5,483	7
Brazil	4,706	6
Belgium-Luxembourg	3,902	5
France	3,791	5
Taiwan	2,521	3
South Korea	2,371	3
Spain	2,306	3
Turkey	1,472	2
Sweden	1,432	2
Denmark	614	1
Other	7,578	9
Total	81,482	100

Source: U.S. Department of Energy, Energy Information Administration, *Quarterly Coal Report* January-April 1985 (Washington, D.C.: Government Printing Office, 1985).

Note: Only countries that received over 2 million tons are listed. Percentages do not add to 100 because of rounding.

Table 35. U.S. Coal Exports to Major Foreign Markets, 1983 (thousands of short tons)

Destination	Metallurgical	Steam	Total
Japan	16,122	1,741	17,863
Canada	6,971	9,838	16,809
Italy	4,970	3,086	8,056
Netherlands	2,251	1,911	4,162
France	2,467	1,676	4,143
Brazil	3,547	—	3,547
Spain	1,761	1,495	3,256
Belgium-Luxembourg	1,840	703	2,543
Taiwan	341	1,535	1,876
Denmark	137	1,602	1,739
South Korea	1,624	44	1,668
Turkey	1,469	159	1,628
Sweden	1,085	492	1,577
Total	44,585	24,282	68,867
Percent of total exports	89	90	90

Source: U.S. Department of Energy, Energy Information Administration, *Weekly Coal Production* (Washington, D.C.: Government Printing Office, 1984).

Note: Only countries receiving over 2 million tons are listed.

Table 36. Bituminous Coal Exports by Customs District, 1983 (thousands of short tons)

Customs District and Port	Total	Metallurgical	Steam
East Coast	44,207	33,887	9,320
Norfolk	36,499	28,557	6,941
Baltimore	6,858	4,712	2,147
Philadelphia	850	618	232
Northern Great Lakes			
Cleveland	16,778	6,941	9,837
Gulf Coast	14,379	8,215	6,164
New Orleans	6,136	1,920	4,216
Mobile	8,243	6,295	1,948
West Coast			
Los Angeles	1,889	736	1,154

Source: Same as table 35.

Table 37. U.S. Metallurgical and Steam Coal Exports by Region of Origin (millions of short tons)

Region	1970	1974	1978	1980	1981	1982	1983	1984
Appalachia	69.6	58.9	39.5	90.7	100.0	98.3	72.3	78.2
Midwest	0.0	0.3	0.1	0.8	6.4	2.8	1.1	1.2
West	0.7	0.6	0.2	1.2	4.9	2.9	2.4	1.4
U.S. Total	70.3	59.8	39.8	92.7	111.3	104.0	75.8	80.7[a]

Source: U.S. Department of Energy, Energy Information Administration, *Coal Distribution* (Washington, D.C.: Government Printing Office, 1970-1984).

[a]U.S. total includes anthracite.

Table 38. Energy Consumption per Unit of Industrial Output (1973 = 100)

Consumer	1960	1973	1974	1975	1976	1977	1978	1979	1980
United States	119	118	94	94	92	86	83	86	85
Japan	111	106	100	101	105	100	95	91	79
Germany	85	101	103	92	89	83	84	85	80
France	112	117	100	91	89	88	89	94	87
United Kingdom	112	99	83	88	89	85	82	81	75
Italy	78	98	98	96	91	88	89	85	81
Canada	115	105	102	100	96	97	93	91	92
Average	105	111	96	94	94	89	86	87	82

Source: OECD, International Energy Agency, *World Energy Outlook* (Paris: OECD/IEA, 1982).

Table 39. Percentage Share of Coal Use in OECD Electricity
Generation and in Industry, 1973, 1978, 1982

Consumer	Electricity Generation			Industry		
	1973	1978	1982	1973	1978	1982
OECD total	36.6	34.4	40.5	20.1	17.8	23.6
North America	43.3	39.4	48.0	18.1	17.1	24.0
Pacific	15.8	15.6	19.2	25.4	21.2	30.5
Europe	38.4	37.8	37.3	20.5	17.0	19.8
Canada	12.9	15.0	18.3	12.0	18.4	21.9
United States	47.2	42.9	52.7	18.7	16.9	24.3
Japan	7.9	5.5	11.0	23.3	18.9	29.3
France	19.9	28.5	23.9	20.2	16.0	16.7
West Germany	64.5	57.5	62.5	25.0	20.5	24.7
Italy	4.5	6.6	13.5	9.0	9.7	12.7
United Kingdom	61.9	65.4	71.8	23.8	18.2	16.1

Source: International Energy Agency, OECD, "1982 Coal Information Report"
(Paris: unpublished report, March 28, 1983); 1982 data from *Coal Information, 1984*,
International Energy Agency, OECD, (Paris: 1984).

Table 40. Coal Consumption by Coke Plants, 1970-1984

Year	Millions of Short Tons	Percent of Total Domestic Coal Consumption
1970	96.5	18.6
1973	94.1	16.7
1974	90.2	16.2
1975	83.6	14.9
1976	84.7	14.0
1977	77.7	12.4
1978	71.4	11.4
1979	77.4	11.4
1980	66.7	9.5
1981	61.0	8.3
1982	40.9	6.8
1983	37.4	5.0
1984	43.8	4.9

Source: U.S. Department of Energy, Energy Information Administration, *1982
Annual Energy Review, Weekly Coal Production, April 1983* (Washington, D.C.:
Government Printing Office, 1983), table 55; *Coal Distribution, 1984*; tables 13 and
16.

Table 41. The Origin of Coking Coal Received by North American Coke Plants, 1970-1984 (millions of short tons)

Year	Kentucky[a]	District 8	District 8 as Percent of Appalachia	Appalachia	Appalachia as Percent of U.S. Total	Total Coal Received by North American Coking Plants
1970	13.2	34.7	40	87.7	85	103.0
1971	11.5	31.0	40	77.7	89	87.2
1972	13.5	36.9	43	86.7	90	96.3
1973	13.9	39.0	42	92.3	94	98.0
1974	14.6	36.7	44	82.7	89	93.2
1975	12.2	34.6	42	81.9	89	92.5
1976	11.7	36.8	44	83.5	90	92.7
1977	10.5	32.6	43	76.3	90	85.0
1978	8.1	29.5	46	64.5	91	71.1
1979	9.9	30.5	40	75.8	90	84.0
1980	8.6	27.8	43	64.8	89	72.7
1981	8.6	27.2	48	56.6	89	63.5
1982	5.2	19.7	50	39.7	76	51.9
1983	5.5	15.5	46	33.9	90	37.4
1984	5.8	18.7	48	39.3	89	43.8

Sources: U.S. Department of Energy, Energy Information Administration, 1977-1980; *Coke and Coal Chemicals*, 1977-1980; *Coal Distribution*, 1977-1984. U.S. Department of the Interior, Bureau of Mines, *Minerals Yearbooks*, "Coal—Bituminous and Lignite," and "Coke and Coal Chemicals, 1970-1976" (Washington, D.C.: Government Printing Office, 1972 to 1985); U.S. Department of Energy, Energy Information Administration, *Quarterly Coal Report, January- March 1985* (Washington, D.C., July 1985).

Note: For 1970 to 1977, destinations of shipments include United States, Canada, Mexico; for 1978 to 1982, United States and Canada only; for 1983 and 1984, United States only.

[a]Represents coal shipped to domestic coke plants only. Also includes an insignificant (less than 1%) amount of coking coal mined in Western Kentucky.

Table 42. Industrial Consumption of Coal by Sector, 1984
(millions of tons)

SIC Sector	Tons Received	Percent of Total
Chemicals and allied products	15,798	24
Stone, clay, glass, and cement	14,847	23
Paper and allied products	11,058	17
Primary metals industries[a]	7,930	12
Food and kindred products	5,084	8
Transportation equipment	2,034	3
Textile mill products	1,719	3
All others	6,471	10
Total	64,706	100

Source: Energy Information Administration, *Quarterly Coal Report, First Quarter* (Washington, D.C.: U.S. Department of Energy, 1985).

[a]Excludes coke plants.

Table 43. Coal Consumption by U.S. Industry, 1970-1984
(millions of short tons)

Year	Consumption by Industry	Shipments of Appalachian Coal to Industry from	
		District 8	Appalachia
1970	90.2	25.6	—
1973	68.0	21.1	40.0[a]
1974	64.9	20.5	41.7
1975	63.6	17.7	34.8
1976	61.8	16.9	32.1
1977	61.5	18.0	35.1
1978	63.1	17.2	34.6
1979	67.7	18.2	38.6
1980[b]	59.4	18.0	35.8
1981[b]	62.1	19.2	36.1
1982[b]	58.8	18.2	32.6
1983[b]	62.8	20.6	36.1
1984[b]	72.1	25.6	42.4

Source: Consumption by industry derived from U.S. Department of Energy, Energy Information Administration, *1982 Annual Energy Review* (Washington, D.C.: U.S. Department of Energy, 1983), p. 127. Shipments data for 1970 to 1975 are from U.S. Department of the Interior, Bureau of Mines, *Mineral Industry Surveys* (Washington, D.C.: Government Printing Office, 1971 to 1975). Data for 1976-1982 are from U.S. Department of Energy, Energy Information Administration, *Coal Distribution* (Washington, D.C.: Government Printing Office, 1977 to 1985).

[a]Exclusive of district 13.
[b]Excludes anthracite.

Table 44. Coal Consumption at Industrial Plants by Selected States, 1984

State	Thousands of Tons
Ohio	5,967
Indiana	5,054
Pennsylvania	4,639
Michigan	4,041
Tennessee	3,758
Virginia	3,563
Illinois	3,325
Alabama	3,046
West Virginia	2,922
New York	2,719
South Carolina	2,226
Kentucky	2,152
Wisconsin	2,019
North Carolina	1,992
Georgia	1,544
Maryland	888
Florida	829
Total	48,684
Percentage of total U.S. consumption	66

Source: U.S. Department of Energy, Energy Information Administration, *Quarterly Coal Report* (Washington, D.C.: Government Printing Office, 1985).

Table 45. Residential and Commercial Consumption of Coal, 1970-1984 (millions of short tons)

Year	Residential and Commercial Consumption
1970	16.1
1973	11.1
1974	11.4
1975	9.4
1976	8.9
1977	9.0
1978	9.5
1979	8.4
1980	6.5
1981	7.4
1982	8.2
1983	8.4
1984	9.1

Source: U.S. Department of Energy, Energy Information Administration, *1982 Annual Energy Review* (Washington, D.C.: U.S. Department of Energy, 1983), p. 127; and U.S. Department of Energy, Energy Information Administration, *Quarterly Coal Report* (Washington, D.C.: Government Printing Office), 1984, p. 26, 1985, p. 32.

Table 46. Production of Bituminous Coal in Appalachia, 1984 (millions of short tons)

State or Area	Output	Percent of Basin Total	Percent of U.S. Total
West Virginia	129.6	30	16
Eastern Kentucky	124.8	28	15
Pennsylvania[a]	71.1	16	9
Ohio	39.0	9	5
Virginia	35.3	8	4
Alabama	27.1	6	3
Tennessee	7.7	2	1
Maryland	4.2	1	1
Total	438.8	100	54

Source: Same as table 13.

[a]Exclusive of anthracite coal.

Table 47. Coal Output for Six Appalachian Coal Basin States, 1970-1984
(millions of short tons)

State or Area	1970	1971	1972	1973	1974	1975	1976	1977	1978	1979	1980	1981	1982	1983	1984	Percentage Change (1970-1984)
West Virginia	144.1	118.3	123.7	115.4	102.5	109.3	108.8	95.4	85.3	112.2	120.3	112.2	128.5	115.0	130.9	− 9.2
Eastern																
Kentucky	72.5	71.6	68.9	74.0	85.4	87.3	91.1	94.0	96.2	104.1	105.6	115.4	111.2	95.6	123.5	+70.3
Pennsylvania	80.5	72.8	75.9	76.4	80.5	84.1	85.8	84.6	81.5	88.9	87.0	78.1	74.8	65.7	72.3	− 10.2
Ohio	55.4	51.4	51.0	45.8	45.4	46.8	46.6	47.9	41.2	43.5	39.2	35.7	36.5	33.7	40.5	− 26.9
Virginia	35.0	30.6	34.0	33.9	34.3	35.5	36.8	37.6	31.9	36.8	41.0	42.0	39.8	35.0	35.4	+1.1
Alabama	20.6	17.9	20.8	19.2	19.8	22.6	24.2	21.5	20.6	24.2	26.4	24.5	26.6	23.8	26.2	+27.2
Total	408.1	362.6	374.3	364.7	367.9	385.6	393.3	381.0	356.7	409.7	419.5	407.9	417.4	383.1	441.2	8.1

Sources: U.S. Department of the Interior, Bureau of Mines, "Coal: Bituminous and Lignite," in *Minerals Yearbook, 1970-1975*, (Washington, D.C.: Government Printing Office). U.S. Department of Energy, Energy Information Administration, *Coal Production, 1976-1981* (Washington, D.C.: Government Printing Office). U.S. Department of Energy, Energy Information Administration, *Quarterly Coal Report* (Washington, D.C.: Government Printing Office, 1985).

Table 48. Production of Coal in Six Appalachian Coal Basin States, Underground and Surface, 1970-1983
(millions of short tons)

State or Area	1970	1971	1972	1973	1974	1975	1976	1977	1978	1979	1980	1981	1982	1983	Percentage Change (1970-1982)
West Virginia															
Underground	116.4	92.4	101.7	95.5	82.2	88.4	87.6	73.5	65.2	91.3	95.8	89.1	100.8	91.9	−21.0
Surface	27.7	25.8	22.1	19.9	20.2	20.9	21.3	21.9	20.1	20.9	24.6	23.1	27.9	22.4	−19.1
Pennsylvania															
Underground	55.4	44.3	49.1	46.2	42.2	44.6	43.8	38.4	32.9	43.0	41.0	34.2	31.2	34.5	−37.7
Surface	25.1	28.5	26.8	30.2	38.2	39.5	42.0	46.3	48.6	45.9	45.9	43.9	43.6	30.6	+21.9
Eastern Kentucky															
Underground	43.2	37.4	37.9	40.6	40.5	40.6	40.6	38.3	41.6	54.1	55.7	59.6	58.7	49.0	+13.4
Surface	29.3	34.2	30.9	33.4	44.8	46.7	50.6	55.7	54.6	49.9	49.9	55.8	52.5	44.1	+50.5
Ohio															
Underground	18.1	12.9	16.3	16.2	14.4	15.5	16.6	14.2	11.9	14.5	12.9	10.7	12.2	10.8	−40.3
Surface	37.2	38.6	34.7	29.6	31.3	31.3	30.0	33.7	29.3	29.1	26.2	25.1	24.9	22.8	−38.7
Virginia															
Underground	28.0	21.6	24.0	23.4	22.8	23.2	26.1	23.1	21.5	27.4	31.4	32.3	31.4	26.8	−4.3
Surface	7.0	9.0	9.0	10.5	11.6	12.3	13.9	14.6	10.4	9.5	9.1	8.3	8.2	7.7	+10.0
Alabama															
Underground	9.1	6.8	7.6	7.6	7.1	7.6	7.4	6.6	6.2	8.4	9.5	8.6	10.6	10.9	+19.8
Surface	11.5	11.1	13.2	11.6	12.8	15.0	14.1	14.9	14.4	15.7	17.5	15.7	14.6	12.8	+11.3

Source: U.S. Department of the Interior, Bureau of Mines, "Coal: Bituminous and Lignite," in Minerals Yearbook, 1969-1975 (Washington, D.C.: Government Printing Office). U.S. Department of Energy, Energy Information Administration, Coal Production, 1976-1983 (Washington, D.C.: Government Printing Office).

aAnthracite excluded.

Table 49. Percentage Distribution of Production in Appalachian Basin, 1970-1983

State or Area	Underground								Surface							
	1970	1975	1978	1979	1980	1981	1982	1983	1970	1975	1978	1979	1980	1981	1982	1983
West Virginia	80.0	80.9	76.4	81.4	79.6	79.4	78.3	80.5	19.2	19.1	23.6	18.6	20.4	20.6	21.7	19.5
Eastern Kentucky	59.6	46.5	43.2	52.0	52.7	51.6	52.9	52.6	40.4	53.5	56.8	48.0	47.3	48.4	47.1	47.4
Pennsylvania	68.8	53.0	40.4	48.4	47.2	43.8	41.7	50.6	31.2	47.0	59.6	51.6	52.8	56.2	58.3	49.4
Ohio	32.7	33.1	28.9	33.3	33.0	29.9	32.9	32.1	67.3	66.9	71.1	66.7	67.0	70.1	67.1	67.9
Virginia	80.0	65.4	67.4	74.3	77.0	79.6	79.3	77.7	20.0	34.6	32.6	25.7	23.0	20.4	20.7	22.3
Alabama	44.2	33.6	30.1	34.9	35.2	35.4	42.1	46.0	55.8	66.4	69.9	65.1	64.8	64.6	57.9	54.0
Appalachia	66.2	57.0	50.3	58.3	58.7	57.7	58.8	60.9	33.8	43.0	49.7	41.7	41.3	42.3	41.2	39.1

Source: Calculated from table 48.

Table 50. Demonstrated Coal Reserve Base of Appalachia, 1974
(millions of short tons)

State or Area	Underground Percentage of Sulfur					Surface Percentage of Sulfur				
	≤1.0	1.1–3.0	>3.0	Unknown	Total	≤1.0	1.1–3.0	>3.0	Unknown	Total
West Virginia	11,087	12,583	6,553	4,143	34,378	3,006	1,423	270	510	5,212
Pennsylvania	7,180	16,195	3,568	2,865	29,819	139	718	232	90	1,181
Ohio	116	5,450	10,109	1,754	17,423	19	991	2,525	118	3,654
Eastern Kentucky	5,043	2,392	213	1,814	9,467	1,516	930	87	915	3,450
Virginia	1,729	945	12	283	2,971	412	218	2	47	679
Alabama	589	1,017	15	176	1,798	35	83	2	1,063	1,184

Source: U.S. Department of the Interior, Bureau of Mines, *Demonstrated Coal Reserve Base, May 1975* (Washington, D.C.: Government Printing Office, 1975).

Note: Totals may not be equal to the sum of components because of rounding.

Table 51. Estimated Production Life of Demonstrated Underground
and Surface Reserves in Appalachia, 1983

	Remaining Years of Production[a]	
State or Area	Underground	Surface
Alabama	79	215
Eastern Kentucky	175	39
Ohio	602	208
Pennsylvania	409	37
Tennessee	72	113
Virginia	45	85
West Virginia	185	182

Source: Calculated from U.S. Department of Energy, Energy Information
Administration, *Coal Production 1983* (Washington, D.C.: Government Printing
Office, 1984).

Note: Estimates are based on calculations derived from table 9 and 1980
production data. They represent years of output at 1980 mining rates.

[a]Calculated using 1983 production figures, a 50% recovery rate for demonstrated
underground reserves, and an 80% recovery ratio for demonstrated surface reserves.

Table 52. Appalachian Coal Output Concentration, 1983
(millions of short tons)

		Percent of Regional Output	
Parent Operating Company	Produced	By Company	Cumulative
Consolidation Coal Co.	27.445	7.3	7.3
A.T. Massey Coal Co.	18.218	4.8	12.1
Island Creek Coal Co.	14.124	3.7	15.8
Pittston Coal Co.	13.473	3.6	19.4
American Electric Power	13.262	3.5	22.9
Eastern Associated Coal Corp.	11.366	3.0	25.9
Bethlehem Mines Corp.	7.854	2.1	28.0
North American Coal Co. (NACCO)	7.409	2.0	30.0
Westmoreland Coal Co.	7.383	1.9	31.9
Total production of largest nine companies	120.534	31.9	31.9
Remainder of Appalachia	257.418	68.1	100.0
Total for region	377.952	100.0	100.0

Source: Calculated from *Keystone Industrial Manual* (New York, N.Y.: McGraw-Hill,
1984).

Table 53. Concentration Ratios for Total U.S. Bituminous Coal Production, by Producing Company or Group

Company Or Group	Percent of Total Production						
	1950	1960	1970	1974	1980	1981	1983
Single largest	4.8	7.0	11.4	11.3	7.2	6.3	6.9
2 largest	9.1	13.9	22.1	19.9	13.2	11.5	12.3
3 largest	11.6	18.2	27.1	23.4	18.1	15.8	17.4
4 largest	13.6	21.3	30.5	26.7	21.5	19.4	21.0
8 largest	19.4	30.4	41.0	44.3	29.5	28.2	30.1
12 largest	23.6	36.5	48.9	50.4	35.4	34.5	37.6
15 largest	26.4	39.5	52.2	54.1	39.9	38.9	42.6
20 largest	30.4	44.4	56.5	58.5	—	—	—
50 largest	45.2	59.9	68.3	72.8	66.3	63.1	70.4[a]
Producers of at least 100,000 tons	82.8	87.0	93.8	94.5	—	—	—
Remaining Companies	17.2	13.0	6.2	5.5	—	—	—

Source: Calculated from *Keystone Coal Industry Manual* (1980-1983) and Phillip E. Griffin, "Industrial Concentration and Firm Diversification in Bituminus Coal with Special Reference to the Southeastern United States, 1950-1970" (Knoxville: Univ. of Tennessee, 1972).

Note: Data include both commercial and captive production. A few enteprises listed as firms might be technically termed as affiliated production groups. Total bituminous coal production in the United States was 516,311,053 tons during 1950; 415,512,347 tons during 1960; 602,932,000 tons during 1970; 603,000,000 tons during 1974; 819,716,000 tons during 1980; 818,352,000 tons during 1981; and 780,000,000 tons during 1983.

[a]Preliminary estimate.

Table 54. Appalachian Basin Coal Mines Producing More Than
1 Million Tons in 1983

	Parent Company	Mine	Location	Tonnage
1.	Central Ohio Coal Co.	Muskingum (S)	Ohio	3,512,135
2.	Consolidation Coal Co.	Loveridge No. 22 (D)	W. Va.	3,425,769
3.	Consolidation Coal Co.	Humphrey No. 7 (D)	W. Va.	2,944,392
4.	Peabody Coal Co.	Sinclar (D & S)	Ky.	2,888,000
5.	Consolidation Coal Co.	Blacksville No. 2 (D)	W. Va.	2,887,947
6.	Enoxy Coal Inc.	Pevler (D & S)	Ky.	2,816,696
7.	MAPCO, Inc.	Martiki (S)	Ky.	2,775,415
8.	Pyro Mining Co.	William Station (D)	Ky.	2,700,000
9.	Peabody Coal Co.	River Queen (S)	Ky.	2,666,000
10.	Eastern Assoc. Coal Corp.	Federal No. 2 (D)	W. Va.	2,605,300
11.	Quarto Mining Co.	Powhatan No. 4 (D)	Ohio	2,556,975
12.	Consolidation Coal Co.	Shoemaker (D)	W. Va.	2,490,000
13.	U.S. Steel Mining Co., Inc.	Pinnacle No. 5 (D)	W. Va.	2,332,343
14.	Southern Ohio Coal Co.	Martinka No. 1 (D)	W. Va.	2,283,031
15.	Greenwich Collieries Div.	North and South (D & S)	Pa.	2,260,371
16.	Consolidation Coal Co.	No. 95 - Robinson Run (D)	W. Va.	2,185,849
17.	Jim Walter Resources, Inc.	Blue Creek No. 3 (D)	Ala.	1,955,000
18.	Jim Walter Resources, Inc.	Blue Creek No. 4 (D)	Ala.	1,876,000
19.	Island Creek Coal Co.	Providence No. 1 (D)	Ky.	1,839,412
20.	Drummond Coal Co.	Cedrum (S)	Ala.	1,806,000
21.	Permac Inc.	Permac No. 3 (D)	Va.	1,800,000
22.	United Coal Co.	Wellmore No. 8 (D)	Va.	1,739,995
23.	Eastern Assoc. Coal Corp.	Kopperston (D)	W. Va.	1,691,189
24.	Drummond Coal Co.	Short Creek (S)	Ala.	1,644,000
25.	Consolidation Coal Co.	Blackville No. 1 (D)	W. Va.	1,541,614
26.	MAPCO, Inc.	Dotiki (D)	Ky.	1,522,784
27.	Southern Ohio Coal Co.	Meigs No. 2 (D)	Ohio	1,502,701
28.	Valley Camp Coal Co.	Donaldson (D)	W. Va.	1,455,000
29.	Consolidation Coal Co.	Pursglove No. 15 (D)	W. Va.	1,451,298
30.	Peabody Coal Co.	Camp No. 1 (D) (D)	Ky.	1,438,000
31.	Cimmaron Coal Corp.	Volunteer (S)	Ky.	1,417,805
32.	Peabody Coal Co.	Sunnyhill (D)	Ohio	1,414,000
33.	Consolidation Coal Co.	Ireland (D)	W. Va.	1,395,000
34.	Drummond Coal Co.	Kellerman (S)	Ala.	1,388,000
35.	United Coal Co.	Wellmore No. 7 (D)	Va.	1,383,661

Table 54 (continued)

	Parent Company	Mine	Location	Tonnage
36.	Florence Mining Co.	Florence No. 1 (D)	Pa.	1,373,304
37.	Consolidation Coal Co.	McElroy (D)	W. Va.	1,323,000
38.	Helvetia Coal Co.	Lucerne No. 6 (D)	Pa.	1,321,326
39.	Consolidation Coal Co.	Arkwright (D)	W. Va.	1,307,455
40.	Hobet Mining & Construction Co., Inc.	Mine No. 21 (S)	W. Va.	1,251,011
41.	U.S. Steel Mining Co., Inc.	Lynch No. 37 (D)	Ky.	1,224,000
42.	Peabody Coal Co.	Camp No. 11 (D)	Ky.	1,219,000
43.	Clinchfield Coal Co.	McClure (D)	Va.	1,205,619
44.	NACCO Mining Co.	Powhatan No. 6 (D)	Ohio	1,182,782
45.	Peabody Coal Co.	Star No. 282 (D)	Ky.	1,175,000
46.	Buffalo Mining Co.	Lorado (D & S)	W. Va.	1,141,072
47.	Bethlehem Mining Corp.	Somerset No. 66 (D)	Pa.	1,140,460
48.	Consolidation Coal Co.	Mohoning Valley No. 33 (S)	Ohio	1,134,453
49.	Enoxy Coal Inc.	Upshur (S)	W. Va.	1,118,455
50.	Peabody Coal Co.	Ken (D & S)	Ky.	1,110,000
51.	MAPCO, Inc.	Mettiki (D)	Md.	1,093,319
52.	Peabody Coal Co.	Alston (S)	Ky.	1,090,000
53.	Pittsburgh & Midway Coal Mining Co.	Colonial (S)	Ky.	1,089,200
54.	Bethlehem Mines Corp.	Ellsworth No. 5 (D)	Pa.	1,082,893
55.	Helen Mining Co.	Homer City (D)	Pa.	1,063,167
56.	Consolidation Coal Co.	Osage No. 3 (D)	W. Va.	1,043,765
57.	Keystone Coal Mining Corp.	Jane (D)	Pa.	1,024,960
58.	MAPCO, Inc.	Retiki (D)	Ky.	1,003,095
	Total			67,619,589

Source: *Keystone Coal Industry Manual,* 1983 (New York, N.Y.: McGraw-Hill, 1984).
Note: D = deep mine; S = surface mine.

Table 55. Size Distribution of Mines Among Appalachian Coal Basin States, 1983 (millions of short tons)

State or Area	Category 1		Category 2		Category 3		Category 4		Category 5		Total	
	No. of Mines	Output	No. of Mines	Output	No. of Mines	Output	No. of Mines	Output	No. of Mines	Output	No. of Mines	Output
Eastern Kentucky												
Underground	10	7.4	41	12.6	80	11.1	141	9.9	301	8.1	573	49.1
Surface	8	12.5	48	14.2	49	6.9	77	5.4	244	5.2	426	44.2
Ohio												
Underground	8	9.8	2	0.7	2	0.3	0	0.0	0	0.0	12	10.8
Surface	9	9.2	19	5.5	21	3.0	44	3.1	80	2.0	173	22.8
Pennsylvania												
Underground	27	24.7	20	6.6	13	1.9	12	0.8	15	0.4	87	34.8
Surface	9	6.3	24	6.8	57	8.1	67	4.9	172	4.4	329	30.5
West Virginia												
Underground	44	52.6	47	14.9	83	11.8	112	8.2	161	4.5	447	91.9
Surface	8	6.6	25	6.8	27	3.9	38	2.6	97	1.0	195	22.4
Virginia												
Underground	7	4.5	14	3.7	49	6.7	96	6.9	167	5.0	333	26.8
Surface	0	0.0	5	1.2	23	3.1	34	2.3	43	1.1	105	7.7
Alabama												
Underground	10	10.3	1	0.4	1	0.1	2	0.1	0	0.0	14	10.9
Surface	5	4.1	11	3.9	19	2.7	15	1.2	34	0.9	84	12.8
Total, all areas	145	148.0	257	77.2	424	59.5	638	45.4	1,314	32.5	2,778	368.1

Source: U.S. Department of Energy, Energy Information Administration, *Coal Production, 1981* (Washington, D.C.: Government Printing Office, 1984).

Categories: 1 = 500,000 tons and over; 2 = 200,000 - 499,999; 3 = 100,000 - 199,999; 4 = 50,000 - 99,999; 5 = 10,000 - 49,999

Table 56. Average Mine Output for Major Appalachian States,
1971, 1976, 1983
(thousands of short tons)

| State or Area | Output per Mine | | | | | |
| | Underground | | | Surface[a] | | |
	1971	1976	1983	1971	1976	1983
Eastern Kentucky	44.468	49.034	85.530	35.002	41.228	103.732
Ohio	367.486	536.323	901.833	144.449	112.620	131.561
Pennsylvania	244.691	312.564	370.202	48.880	59.850	87.336
West Virginia	148.136	106.910	205.633	60.613	61.132	114.723
Alabama	421.938	352.667	775.714	113.071	59.877	151.964
Virginiai	60.932	65.301	80.417	28.562	35.205	73.695

Source: U.S. Department of the Interior, Bureau of Mines, *Coal: Bituminous and Lignite, 1971*; U.S. Department of Energy, Energy Information Administration, *Coal: Bituminous and Lignite in 1976*; U.S. Department of Energy, Energy Information Administration, *Coal Production, 1983* (Washington, D.C.: Government Printing Office, 1971, 1976, 1984).

[a]Includes auger mining.

Table 57. Percentage Distribution of Mines and Output by Size for Appalachian Coal Basin States, 1983

State or Area	Category 1		Category 2		Category 3		Category 4		Category 5	
	Percent of Mines	Percent of Output	Percent of Mines	Percent of Output	Percent of Mines	Percent of Output	Percent of Mines	Percent of Output	Percent of Mines	Percent of Output
Eastern Kentucky										
Underground	1.7	14.9	7.2	25.7	14.0	22.4	24.6	20.2	52.5	16.3
Surface	1.8	28.3	11.3	31.9	11.5	15.6	18.1	12.2	57.3	11.8
Ohio										
Underground	66.7	91.6	16.7	6.5	16.7	2.8	0.0	0.0	0.0	0.0
Surface	5.2	40.4	11.0	24.1	12.1	13.2	25.4	13.6	46.2	8.8
Pennsylvania										
Underground	28.7	71.0	21.3	19.0	13.8	5.4	12.8	2.3	16.0	1.1
Surface	2.7	20.7	7.3	22.3	17.2	26.6	20.4	16.1	52.3	14.4
West Virginia										
Underground	9.8	57.2	10.5	16.2	18.6	12.8	25.0	8.9	36.0	4.9
Surface	4.1	29.5	12.8	30.4	13.8	17.4	19.5	11.6	49.7	4.5
Virginia										
Underground	2.1	16.8	4.2	13.8	14.7	25.0	28.8	25.7	50.1	18.7
Surface	0.0	0.0	4.8	15.6	21.9	40.3	32.4	29.9	41.0	·14.3
Alabama										
Underground	71.4	94.5	7.1	3.7	7.1	0.9	14.3	0.9	0.0	0.0
Surface	6.0	32.0	13.1	30.5	22.6	21.1	17.9	9.4	40.5	7.0

Source: Calculated from U.S. Department of Energy, Energy Information Administration, *Coal Production, 1983* (Washington, D.C.: Government Printing Office, 1984).

Table 58. Productivity in Appalachian Bituminous Coal Mining,
1982, 1983
(average short tons per miner per day)

State or Area	Underground		Surface	
	1982	1983	1982	1983
Eastern Kentucky	12.10	14.29	21.49	20.69
West Virginia	10.51	13.39	16.83	19.70
Pennsylvania	9.15	10.64	14.82	18.77
Ohio	8.78	10.68	19.63	20.79
Appalachia[a]	10.48	12.49	18.04	19.26

Source: Energy Information Administration, Department of Energy, *Coal Production, 1982, 1983* (Washington, D.C., 1983, 1984).

[a]Included are also Alabama, Maryland, Virginia, and Tennessee.

Table 59. Average Number of Days Worked by Coal Miners in the
Four Major Appalachian Coal-Producing States, 1970-1983

State or Area	1970	1975	1978	1979	1980	1981	1982	1983
Underground								
Eastern Kentucky	189	231	161	183	194	181	178	184
West Virginia	225	221	131	199	199	176	179	179
Pennsylvania	250	247	184	229	215	198	199	214
Ohio	233	216	154	219	215	189	195	184
Surface								
Eastern Kentucky	163	210	138	164	187	188	175	185
West Virginia	185	212	160	197	208	178	190	188
Pennsylvania	246	240	197	231	231	228	225	220
Ohio	242	256	197	243	230	221	221	208

Source: U.S. Department of Energy, Energy Information Administration, *Coal Production, 1978, 1980, 1981, 1982, 1983* (Washington, D.C.: Government Printing Office, 1978, 1980, 1981); U.S. Department of the Interior, Bureau of Mines and Minerals, *Coal: Bituminous and Lignite, 1975, 1970* (Washington, D.C.: Government Printing Office, 1975, 1970).

Table 60. Price per Ton of Coal, 1973-1983

State or Area	Underground				Surface			
	Price in Current Dollars	Current Price Ratio 1983 to 1973	Price in 1973 Dollars	Constant Price Ratio, 1983 to 1973	Price in Current Dollars	Current Price Ratio, 1983 to 1973	Price in 1973 Dollars	Constant Price Ratio, 1983 to 1973
Eastern Kentucky								
1973	10.63	1.00	10.63	1.00	7.05	1.00	7.05	1.00
1975	27.03		22.72		15.24		12.81	
1978	28.86		20.29		22.58		15.87	
1980	30.98		18.36		26.23		15.55	
1981	32.47		17.60		28.86		15.64	
1982	32.71		16.72		28.85		14.75	
1983	30.71	2.89	15.20	1.43	28.43	4.03	14.10	2.00
West Virginia								
1973	12.24	1.00	12.24	1.00	8.58	1.00	8.58	1.00
1975	30.60		25.73		25.52		21.45	
1978	35.45		24.92		25.70		18.07	
1980	36.46		21.61		28.72		17.02	
1981	39.38		21.34		32.94		17.85	
1982	38.53		19.70		34.31		17.54	
1983	36.45	2.98	18.04	1.47	31.36	3.66	15.52	1.81
Pennsylvania								
1973	12.02	1.00	12.02	1.00	7.68	1.00	7.68	1.00
1975	30.41		25.57		19.09		16.05	
1978	37.75		26.54		21.15		14.87	
1980	37.61		22.29		23.84		14.13	
1981	36.51		19.79		26.99		14.63	

1982	37.46		19.15		30.46		14.57	
1983	37.35	3.11	18.49	1.54	30.19	3.93	14.95	1.95
Ohio								
1973	8.50	1.00	8.50	1.00	6.82	1.00	6.82	1.00
1975	18.76		15.77		15.57		13.09	
1978	31.50		22.15		17.82		12.53	
1980	38.42		22.77		22.60		13.39	
1981	37.86		20.52		25.39		13.76	
1982	42.36		21.65		26.95		13.78	
1983	44.68	5.26	22.12	2.60	28.00	4.11	13.86	2.03
Virginia								
1973	12.70	1.00	12.70	1.00	7.66	1.00	7.66	1.00
1975	34.44		28.95		23.72		19.94	
1978	33.05		23.24		25.24		17.74	
1980	36.47		21.62		28.05		16.63	
1981	36.19		19.61		30.18		16.36	
1982	35.66		18.23		30.30		15.49	
1983	32.71	2.58	16.19	1.28	28.94	3.78	14.47	1.89
Alabama								
1973	15.54	1.00	15.54	1.00	8.04	1.00	8.04	1.00
1975	33.77		28.39		22.87		19.23	
1978	39.22		27.57		27.94		19.64	
1980	43.37		25.71		33.56		19.89	
1981	50.40		27.31		36.67		19.87	
1982	49.56		24.33		38.45		19.65	
1983	45.58	2.93	22.56	1.45	38.94	4.84	19.27	2.40

Source: Same as table 59.

Note: Constant 1973 prices were calculated using the implicit price deflator for gross national product.

Table 61. Productivity and Price for Selected
Coal-Producing States, 1983

State or Area	Tons Mined per Labor Day	Real Price per Ton (1973 dollars)
Underground mines		
Eastern Kentucky	14.29	$15.20
West Virginia	13.39	18.04
Pennsylvania	10.40	18.49
Ohio	10.68	22.12
Surface mines		
Eastern Kentucky	20.69	$14.10
West Virginia	19.70	15.52
Pennsylvania	16.65	14.95
Ohio	20.79	13.86

Source: Tables 59 and 60.

Table 62. Percent of Coal Sold in Open and Captive Markets
and Average Mine Price, 1970-1983

State or Area	Percent of Output Sold in Market		Average Value per Ton ($)	
	Open	Captive	Open	Captive
Eastern Kentucky				
1973	89	11	8.44	13.72
1975	a	a	18.38	41.10
1978	93	7	a	a
1980	89	11	27.90	35.78
1981	90	10	30.12	36.20
1982	88	12	30.54	33.35
1983	93	7	29.38	32.83
West Virginia				
1973	89	11	11.13	15.62
1975	87	13	28.04	38.31
1978	84	16	31.12	44.13
1980	79	21	32.56	43.53
1981	80	20	35.86	46.96
1982	85	15	36.54	44.54
1983	86	14	34.43	41.88
Pennsylvania				
1973	69	31	8.71	13.88
1975	71	29	20.66	35.68
1978	79	21	22.41	48.47
1980	78	22	27.14	41.90
1981	80	20	31.05	41.66
1982	82	18	31.19	40.32
1983	79	21	32.28	39.70
Ohio				
1973	89	11	7.30	8.22
1975	a	a	16.34	a
1978	88	12	a	a
1980	81	19	26.82	32.06
1981	82	18	27.92	34.51
1982	81	19	30.78	37.79
1983	80	20	31.17	42.23

Source: U.S. Department of the Interior, Bureau of Mines, *Coal: Bituminous and Lignite, 1973* and *1975*; U.S. Department of Energy, Energy Information Administration, *Coal Production, 1978–1983* (Washington, D.C.: Government Printing Office, 1975, 1977, 1984).

[a]Withheld to avoid disclosure of individual company data.

Table 63. Sulfur Dioxide Emissions by Source, 1980
(percentages)

	Emissions	
Source	By Fuel	All Fuels
Utility fuel combustion		61.7
Coal	55.4	
Oil	6.2	
Industrial production processes		18.2
Industrial fuel combustion		13.4
Coal	6.5	
Oil	5.9	
Transportation		3.1
Commercial-institutional		2.6
All other		1.2
Total		100.2[a]

Source: U.S. Department of Energy, Energy Information Administration, *Impacts of the Proposed Clean Air Act Amendments of 1982 on the Coal and Electric Utility Industries* (Washington, D.C.: Government Printing Office, 1983).

[a]Because of rounding, does not add to 100.

Table 64. U.S. Regional Coal Production Using Constant Factor Prices,
1.2 lb./MMBtu Sulfur Standard
(millions of short tons)

Year	Northern Appalachia[a]	Southern Appalachia[b]	Midwest[c]	Montana-Wyoming	Utah-Colorado	Arizona-New Mexico	Total
1980	201.7	247.6	134.7	193.8	21.1	54.2	853.1
1985	161.4	317.0	79.1	349.9	32.3	81.2	1,020.9
1990	342.1	198.5	148.4	616.4	100.1	76.8	1,482.3
1995	338.5	165.8	159.6	1,009.5	52.8	49.7	1,775.9
2000	200.7	216.4	308.7	1,077.9	84.8	45.5	1,934.0

Source: Derived from M. Zimmerman, *The U.S. Coal Industry: The Economics of Policy Choice* (Cambridge, Mass.: MIT Press, 1981), p. 71.

[a]Includes Pennsylvania, northern West Virginia, Ohio, and Maryland.
[b]Includes southern West Virginia, Eastern Kentucky, Virginia, Tennessee, and Alabama.
[c]Includes Illinois, Indiana, and Western Kentucky.

Table 65. Strippable Reserves and Slope Angle

	Total Reserves (millions of short tons)[a]	Percentage of Reserves at Slope Angle, in Degrees			
		⟨14.9	15-19.9	20-24.9	⟩25
Alabama	3,223.5	83.1	7.8	5.9	3.2
Eastern Kentucky	4,204.2	10.9	13.9	28.6	46.6
Maryland	109.7	98.6	1.0	0.4	0.0
Ohio	5,989.4	91.3	7.7	1.0	0.0
Pennsylvania	1,297.1	98.8	1.0	0.1	0.1
Tennessee	327.5	62.2	16.4	17.9	3.5
Virginia	846.2	0.0	14.1	58.1	27.8
West Virginia	5,195.1	38.2	18.8	24.3	18.7

Source: Calculated from U.S. Senate, 93rd Cong., Committee on Interior and Insular Affairs, *Coal Surface Mining and Reclamation* (Washington, D.C.: Government Printing Office, 1983), p. 4.

[a]See table 10.

Table 66. Costs of Externalities and Costs of Internalizing Them by Regulation (per ton of coal)

	Appalachia	Midwest	West
Internatization costs			
Due to state regulations	$0.70	$0.51	$0.13
Due to the SMCRA	5.44	2.56	0.41
Total	$6.14	$3.07	$0.54
Externality costs			
Water degradation	$0.10	$0.06	$0.01
Land degradation	1.42	0.75	0.01
Flooding	0.10	—	—
Reduced recreation	0.53	0.22	0.04
Reduced aesthetics—local	1.08	0.31	0.04
Reduced aesthetics—outsiders	1.26	2.00	0.93
Total	$4.49	$3.34	$1.03

Source: Internalization costs are calculated from C.E. Man and J.N. Heller, *Coal and Profitability: An Investor's Guide* (New York: McGraw-Hill, 1979). Externality costs are from J.P. Kalt, "Energy Supply, Environmental Protection, and Income Redistribution under Federal Regulation of Coal Strip Mining," Studies in Energy and the American Economy, Discussion Paper no. 14 (Cambridge: Harvard Univ., January 1982).

Note: Costs are in 1981 dollars.

Table 67. U.S. Fatality Rates in Coal Mining

Time Period	Mean Fatality Rates
1932–1941	1.49
1942–1952	1.14
1953–1969	1.05
1970–1980	0.48

Source: U.S. Department of the Interior, Bureau of Mines, Mine Safety and Health Administration, *Injury Experience in Coal Mining* (Washington, D.C.: Government Printing Office, 1980); Charles S. Perry, "Safety Laws and Spending Save Lives" (Lexington: Univ. of Kentucky, Department of Sociology, December 1981).

Table 68. Tons of Appalachian Coal Mined Per Fatality (millions of short tons)

Years	Underground	Surface
1975–1978	2.55	9.76
1970–1974	2.04	11.04
1965–1969	1.60	9.00
1960–1964	1.20	8.16
1955–1959	0.97	6.50
1950–1954	0.83	3.80
1945–1949	0.59	2.87
1940–1944	0.44	1.88
1935–1939	0.39	1.88
1930–1934	0.35	1.32

Source: C.S. Perry and Christian Ritter, "Dying to Dig Coal, Fatalities in Deep and Surface Coal Mining in Appalachia States," Working Paper RS68 (Lexington: Univ. of Kentucky, Department of Sociology, December 1981).

Table 69. Coal Transport and Its Share in Total Railroad Tonnage,
1970-1983

Year	Millions of Short Tons of Coal		Percentages	
	Total Production	Transported by Rail[a]	Coal Moved by Rail/Total Production	Coal Tons Shipped/Total Tons Shipped by Rail
1970	602.9	425.9	70.6	26.9
1975	648.4	429.9	66.3	29.1
1978	665.1	404.5	60.8	27.4
1979	776.3	500.1	64.4	32.4
1980	823.6	551.4	66.9	35.4
1981	818.4	554.0	67.7	36.6
1982[b]	813.7	503.0	61.8	NA
1983[b]	784.6	470.4	60.0	NA

Source: U.S. Department of Energy, Energy Information Agency publications, *Railroad Deregulations: Impact on Coal; Coal Distribution, 1982–1983; and Quarterly Coal Report* (Washington, D.C.: Government Printing Office, 1983, 1984, and 1985, respectively).

[a]Includes coal transported by rail for domestic consumption, to Canada, and to overseas export ports.

[b]1982 and 1983 data are not completely comparable to previous years, because of different sources of origin.

Table 70. Selected Barge Rates in 1982 Dollars per Ton

| | Port of Departure | | | | | |
| | Mobile | | | New Orleans | | |
Origin	Low	Average	High	Low	Average	High
Northern Appalachia	—	—	—	12.50	15.50	20.00
Central Appalachia	12.00	15.00	20.00	5.50	7.50	9.00
Southern Appalachia	3.00	4.00	6.50	4.00	6.00	8.00

Source: Appalachian Regional Commission, Washington, D.C., November 1982.

Table 71. Bituminous Coal Exports, by Seaport
(thousands of short tons)

Port	1979	1980	1981	1982	1983	1984[a]
Hampton Roads	33,755	51,773	52,041	52,677	35,499	35,630
Baltimore	9,139	12,127	12,868	11,747	6,859	7,195
New Orleans/Baton Rouge	1,410	3,826	13,902	7,674	6,149	5,505
Mobile	1,285	2,445	3,522	9,412	8,243	8,852
Los Angeles/Long Beach	13	1,038	5,262	2,833	1,767	1,451
Philadelphia	55	1,522	2,463	1,516	1,282	1,152
Other	242	382	2,455	1,282	1,097	1,507
Total	45,899	73,113	92,513	87,141	60,896	61,292
Great Lakes	18,885	16,769	17,731	18,104	16,876	20,191
Grand total	64,783	89,882	110,244	105,245	77,772	81,483

Source: U.S. Department of Commerce, Bureau of the Census, *Monthly Report,* EM 522 (Washington, D.C.: Government Printing Office); Energy Information Agency, Department of Energy, *Quarterly Coal Reports,* 1983, 1984 (Washington, D.C.: U.S. Government Printing Office).

[a]Preliminary estimates.

Table 72. Estimated International Fossil Reserves, 1980

Region	Coal[a] Total Recoverable (billions of short tons)	Crude Oil[b] (billions of barrels)	Natural Gas[b] (trillions of cubic feet)
North America			
Canada	6.5	7.1	91
Mexico	—	48.6	77
United States	283.4	27.3	198
Total	289.9	83.0	367
Central and South America			
Argentina	—	2.3	25
Brazil	—	2.0	3
Ecuador	—	1.4	3
Venezuela	—	25.8	55
Other	19.3	3.2	22
Total	19.3	34.7	108
Western Europe			
Netherlands	—	0.3	60
Norway	—	8.3	89
United Kingdom	5.1	13.6	28
West Germany	71.7	0.3	6
Yugoslavia	18.2	—	—
Other	7.6	1.9	16
Total	102.6	24.4	207
Eastern Europe and U.S.S.R.			
U.S.S.R.	264.9	63.0	1,450
Poland	43.2	—	—
Other	14.8	2.0	17
Total	322.9	65.0	1,467
Africa			
Algeria	—	9.0	109
Libya	—	21.1	21
Nigeria	—	16.7	36
South Africa	57.0	—	—
Other	8.2	8.7	21
Total	65.2	55.5	187
Middle East			
Iran	—	48.5	479
Iraq	—	44.5	29
Kuwait	—	92.7	37
Oman	—	3.5	7
Qatar	—	3.4	150

Table 72 (continued)

Region	Coal[a] Total Recoverable (billions of short tons)	Crude Oil[b] (billions of barrels)	Natural Gas[b] (trillions of cubic feet)
Middle East (continued)			
Saudi Arabia	—	171.7	127
United Arab Emirates	—	32.5	32
Other	—	1.6	7
Total		398.4	869
Far East and Oceania			
Australia	72.4	1.4	18
China	108.9	19.1	31
Indonesia	—	8.7	40
Malaysia	—	3.0	50
Other	5.4	5.4	38
Total	186.7	37.6	197

Source: U.S. Department of Energy, Energy Information Administration, *Annual Energy Review, 1984* (Washington, D.C.: Government Printing Office, 1985).

[a]1980 data; includes anthracite, bituminous and subbituminous, and lignite. At least one-half of the in-place known coal is estimated as recoverable.

[b]All reserve figures, except those for U.S.S.R. and gas reserves in Canada, are proven reserves recoverable with present technology and under present prices. U.S.S.R. figures

Table 73. Average Annual Growth in Electric Power Generation and the Share of Coal as a Primary Fuel, 1982-2000

Region	1982–1990 Percent Growth	1982–1990 Percent Coal Based	1991–2000 Percent Growth	1991–2000 Percent Coal Based
North America	2.6	48–53	2.5	50
Europe	2.6	42–45	2.6	43
Pacific	3.4	23–27	2.9	32

Source: Energy balances of OECD countries (1982) and country submissions (1990 and 2000), as shown in *Coal Prospects and Policies, 1983 Review*, International Energy Agency (Paris: OECD, 1984).

Table 74. Estimated Costs of Generating Electricity, by Fuel
(millions of 1981 dollars per kilowatt hour)

| | Oil (2 x 600 MW) | | Nuclear Power (2 x 100 MW) | Coal (with FGD, 2 x 600 MW) | | |
	Low Sulfur	High Sulfur		U.S.	Europe	Japan
Capital cost	16.5	19.8	53.8	27.0	27.0	28.0
Operating cost	2.5	4.2	4.2	5.1	5.1	5.1
Fuel cost	54.6	47.6	10.0	16.0	26.0	26.0
Total cost	73.6	71.6	68.0	48.1	58.1	59.1
Capital investment in dollars per kilowatt produced at:						
10% interest	577	692	1,331	920	920	956
15% interest	617	740	2,011	1,005	1,005	1,045

Source: International Energy Agency, *World Energy Outlook* (Paris: OECD, 1982).

Note: A capacity factor of 65% was used as the current average operation of all power plants. The costs of removing SO_2, NO_x, dust, particulates, and ash to meet legal requirements are included.

Table 75. Comparison of
Domestic Coal Consumption and Export Forecasts
(millions of short tons)

Markets and Predictors	1984	1985	1990	Percentage Growth Rate 1983-1990
Utility market				
(1983 actual consumption—625 million tons)				
EIA	672	679	785	3.31
ICF	646	691	791	3.42
DRI	642	656	704	1.71
Chase	625	649	757	2.78
NCA	618	633	681	1.23
Range	618–663	633–691	681–791	
Average	639	662	744	2.52
Industrial retail market				
(1983 actual consumption—74 million tons)				
EIA	78	79	92	3.16
ICF	76	79	109	5.69
DRI	77	80	100	4.40
Chase	87	89	85	2.00

Table 75 (continued)

Markets and Predictors	1984	1985	1990	Percentage Growth Rate 1983-1990
Industrial retail market (continued)				
NCA	75	79	96	3.79
Range	75–87	79–89	85–109	
Average	79	81	96	3.79
Metallurgical market				
(1983 actual comsumption—37 million tons)				
EIA	46	48	51	4.69
ICF	54	60	63	7.90
DRI	50	52	57	6.37
Chase	42	44	51	4.69
NCA	45	48	50	4.39
Range	42–54	44–60	51–63	
Average	47	50	54	5.55
Export market				
(1983 actual consumption—78 million tons)				
EIA	77	83	105	4.34
ICF	78	80	74	− 7.5
DRI	88	98	115	5.70
Chase	78	76	87	1.57
NCA	79	89	113	5.44
Range	77–88	76–98	74–115	
Average	80	85	99	3.46
Totals				
(1983 actual consumption—814 million tons)				
EIA	873	890	1033	3.46
ICF	854	910	1037	3.50
DRI	857	886	976	2.63
Chase	832	858	980	2.69
NCA	817	849	940	2.09
Range	817–873	849–910	940–1037	
Average	847	879	993	2.88

Source: U.S. Department of Energy, Energy Information Administration (EIA), *Annual Energy Outlook 1983,* May 1984; ICF, *Quarterly Coal Reports; Coal Market Assessment, Fall-Winter 1983* (Washington, D.C.: February 1984); Data Resources Inc. (DRI), *Energy Review, Winter 1983–1984* (Lexington, Mass.: DRI, January 1984); Chase Econometrics, *Chase Energy Analysis Quarterly, Fourth Quarter 1983,* (Bala Cynwyd, Pa.: January 1984); National Coal Association, *Long-Range Forecast for U.S. Coal, Coal Markets in the Future* (Washington, D.C.: March 1984); Nerco Coal Co., *General Marketing Trends and Forecasting,* (Cincinnati, Ohio: July 1984).

NOTES

I COAL: AN OVERVIEW

1. The British thermal unit (Btu) is a measurement of the amount of heat (calories) required to raise 1 pound of water 1° F. at sea level. The caloric content of coal is expressed as Btu per pound.

2. Geological resources are defined as resources that may become economically feasible to recover in the future. Technically and economically recoverable reserves are coal deposits recoverable under prevailing conditions.

3. Sales of electric energy to all ultimate customers—residential, commercial, and industrial—increased in 1981 by 52,654 gigawatt hours, or 2.5%. Only in 1984 did electric energy sales rebound significantly, supported by a burgeoning economy. U.S. Department of Energy, Energy Information Administration, *Electric Power Annual, 1984* (Washington, D.C.: Government Printing Office), p. 123.

II PRINCIPLES OF ENERGY RESOURCE ALLOCATION

1. Expressed algebraically, this relationship can be written $D_c = k - lP_c$, where k is a parameter describing the amount of coal that would be demanded at a price of zero, l is a constant that relates change in quantity demanded to change in price, p_c is the price of coal, and D_c is the quantity of coal demanded.

2. This concept can be expressed algebraically:
$n = -\%\Delta Q/\%\Delta P = -[(\Delta Q)/Q] / [(\Delta P)/P] = -[(\Delta Q)/\Delta P] / (P/Q)$ where Δ is "the change in" and P and Q denote price and quantity demanded, respectively. Unfortunately, data problems make it difficult to estimate such coefficients exactly. Most elasticity coefficient estimates should be regarded as statistically accurate only for the data at hand under the stated assumptions.

3. See H. Houthakker and L. Taylor, *Consumer Demand in the U.S., 1929-1970* (Cambridge, Mass.: Harvard Univ. Press, 1966); F. Bell, "The Pope and the Price of Fish," *American Economic Review*, December 1968.

4. In technical terms, this is referred to as zero or low elasticity of technical substitution: zero if no substitution takes place, and low if the elasticity coefficient is between zero and 1.

5. In technical terms, this situation would be an illustration of an elastic technical substitution. The elasticity coefficient is greater than 1.

6. If a change in the price of an energy resource leaves unaffected the demand for it, then the price elasticity of demand is said to be zero; demand is price inelastic, an extreme case. If the elasticity coefficient falls between zero and 1, relative inelasticity

is said to exist. For a coefficient of 1, unitary elasticity applies; the change in quantity demanded is proportionate to the change in price. Above 1, price elasticity is said to exist.

7. Such conservation will be implemented only if existing energy use is thermodynamically inefficient. See E.R. Berndt and D. Woods, "Technology, Price, and the Derived Demand for Energy," *Review of Economics and Statistics* vol. 57 (Aug. 1975).

8. M.M. Ali, C.E. Harvey, and J.F. Stewart, *Regional Demand and Supply Behavior by Sectors of the U.S. Coal Industry* (Lexington: Univ. of Kentucky, Institute for Mining and Minerals Research, 1981).

9. Most recently, in a yet unpublished paper dated November 1984, Professor A. Goodman, Johns Hopkins University, estimated price elasticities of demand for metallurgical coal for eleven European countries for the period 1978-1981. Using pooled cross-section time series samples, Goodman found the elasticities varying from -0.09 to 0.20. These results, consistent with similar findings by others, suggest that at this time no good substitute exists for coal in the production of coke.

10. Expressed algebraically, this relationship is written $S_c = f(P_c)$, $S_c = m + nP$, where m is a parameter describing the amount of the good that would be offered at a zero price and n denotes a constant that relates changes in quantity supplied to changes in price.

11. If the elasticity coefficient exceeds 1, supply is said to be elastic; if the coefficient is between zero and 1, it is relatively inelastic (or of low elasticity); if the coefficient is 0, supply is said to be completely unresponsive to changes in price.

12. See L. White, "Searching for the Critical Industrial Concentration Ratio," in *Studies in Nonlinear Estimation*, ed. S. Goldfeld and R. Quandt (Cambridge, Mass.: Ballinger, 1976).

13. R.E. Shrieves, "Geographic Market Areas and Market Structure in the Bituminous Coal Industry," *Anti-Trust Bulletin* vol. 23, no. 3 (Fall 1978).

14. H.R. 8619, Pub. Law 93-135, 93 Cong., 1 sess., 1973.

15. See J.V. Koch, *Industrial Organization and Prices* (Englewood Cliffs, N.J.: Prentice-Hall, 1974), p. 78.

16. See J.P. Kalt, *Energy Supply, Environmental Protection, and Income Redistribution under Federal Regulation of Coal Strip Mining*, MIT-EL 81-066 (Cambridge, Mass.: Massachusetts Institute of Technology, Energy Laboratory, 1982).

III DEMAND FOR APPALACHIAN COAL

1. U.S. Federal Trade Commission, *In the Matter of Kennecott Copper Corp., Findings of Fact, Conclusion and Final Order*, Docket 8765 (Washington, D.C.: May 5, 1971), p. 5.

2. G. Stigler, *The Theory of Price* (New York: Macmillan, 1966), p. 85.

3. R.E. Shrieves, "Geographic Market Areas," pp. 597-603.

4. U.S. Department of Energy, Energy Information Administration, *Monthly Energy Review* (March 1985), p. 80.

5. U.S. Department of Energy, Energy Information Administration, *Coal Distribution, January-December 1982* (Washington, D.C.: Government Printing Office, 1983), p. 23.

6. See E. Erickson and R. Spann, "Supply Response in a Regulated Industry: The Case of Natural Gas," *Bell Journal of Economics and Management Science* vol. 2, no. 1 (Spring 1971), pp. 94-121.

7. U.S. Department of Energy, Energy Information Administration, *1982 Annual Energy Review* (Washington, D.C.: Government Printing Office, 1983); *Monthly Energy Review*, March 1985 (Washington, D.C.: Government Printing Office, 1985).

8. Battelle Memorial Institute, *Energy Perspectives* (Columbus, Ohio: Battelle, 1973), p. 3.

9. In Tennessee it is the Tennessee Valley Authority that produces nuclear power.

10. C.E. Harvey, *The Economics of Kentucky Coal* (Lexington: Univ. Press of Kentucky, 1977), p. 32.

11. Kentucky Energy Cabinet, Division of Coal Development, "Kentucky Coal Transportation Study," Interim Report no. 2 (prepared for Kentucky Department of Energy, 1983).

12. In 1980, the delivered price of South African coal (which typically contains 1.13 pounds of SO_2 per million Btu produced) in Florida was $1.83 per million Btu; that of Eastern Kentucky coal (which contains 2.61 pounds of SO_2 per million Btu produced) was $2.10 per million Btu; that of Polish coal (2.19 pounds of SO_2 per million Btu) was $1.69 per million Btu.

13. In 1981 alone, Polish coal production fell 14%, and exports fell by one-half.

14. For example, despite government subsidies equivalent to $823 million, the British National Coal Board reported losses of $158 million in the year ended March 31, 1983 (pounds converted at the June 24, 1983, exchange rate).

15. This district, which includes Eastern Kentucky and parts of West Virginia, Virginia, and Tennessee is also frequently referred to as central Appalachia.

16. Organization for Economic Cooperation and Development, International Energy Agency, *World Energy Outlook* (Paris: OECD/IEA, 1982), p. 105.

17. Cross-price elasticities of demand with natural gas and oil were even lower. See R.S. Pindyck, *The Structure of World Energy Demand* (Cambridge, Mass.: MIT Press, 1979), p. 222.

18. U.S. Department of Energy, Energy Information Administration, *Interim Report of the Interagency Task Force* (Washington, D.C.: Government Printing Office, 1981).

19. OECD/IEA, *World Energy Outlook*, p. 106.

IV SUPPLY OF APPALACHIAN COAL

1. Harvey, *Kentucky Coal*, p. 50.

2. Ibid., p. 51.

3. R. Caves, *American Industry: Structure, Conduct, Performance* (Englewood Cliffs, N.J.: Prentice-Hall, 1982), p. 70, has stated that "probably four-firm concentration has to reach about 50% (roughly equivalent to eight-firm concentration around 70%) before rival firms begin to recognize their oligopolistic interdependence significantly."

4. See Harvey, *Kentucky Coal*, p. 69.

5. "The decline in employee productivity is almost entirely due to the fact that nonproducing workers have had to be hired." A. Schweitzer, *The Limits of Kentucky Coal Output: A Short-Term Analysis* (Lexington: Univ. of Kentucky, Institute for Mining and Minerals Research, 1973), p. 7.

6. R.S. Gordon, "A Reinterpretation of the Pure Theory of Exhaustion," *Journal of Political Economy* vol. 75, no. 3 (June 1967), pp. 274-308; R.S. Pindyck, "Uncertainty and Exhaustible-Resource Markets," *Journal of Political Economy* vol. 88, no.

6 (1980), pp. 1203-25; and Pindyck, "The Optimal Production of an Exhaustible Resource when Price Is Exogenous and Stochastic,'" *Scandinavian Journal of Economics* vol. 83, no. 2 (1981), pp. 277-88.

7. M.E. Slade, "Grade Selection under Uncertainty: Least-Cost Last and Other Anomalies," Discussion Paper 84-01 (Vancouver: Univ. of British Columbia, Department of Economics, Jan. 1984).

8. See M.E. Greenbaum, *Kentucky Coal Reserves: Effects of Coal Industry Structure and Output* (Lexington: Univ. of Kentucky, Institute for Mining and Minerals Research, 1975).

9. For all regressions, productivity is measured as output per day of labor and mine size is measured as output per annum. This may lead to distortion in the results; unfortunately, no other measure of productivity exists.

10. The price per ton of coal is determined graphically as

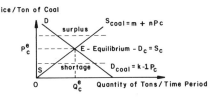

The above is a conventional market diagram. The notations S_c and D_c stand for the supply of and demand for coal; m and k show the amount of coal that would be supplied and demanded at a zero price, and n and j are constants which relate changes in quantities supplied and demanded to changes in price.

11. The time link between spot and contract prices may be illustrated in the following manner. As table 61 shows, the real average price of most Appalachian coal has been declining since 1975, but the spot price has declined more rapidly. Over time, as new contracts are negotiated while prices are falling, long-run prices are being pulled down as well, albeit more gradually. Thus, steadily falling spot prices make it more attractive for coal buyers to purchase more coal on the spot market and less on contract, and prices of the latter begin a gradual decline.

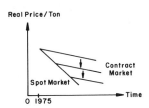

12. There are basically four contract types, all of which allow foreseeable cost increases to be passed forward in the form of higher prices: (1) base price plus escalator clause, (2) cost plus return contract, (3) periodic pricing or renegotiation contract, and (4) market price reopener contract.

13. Charles River Associates, *Coal Price Formation* (Palo Alto, Calif.: Electric Power Research Institute, 1977), pp. 3-16.

14. A modest price increase occurred in Pennsylvania and a sharper one occurred in Ohio, but both were short lived.

V REGULATION IN THE COAL INDUSTRY

1. C.E. Harvey, *Financing Public Expenditures for Energy-Impacted Roads* (Lexington: Univ. of Kentucky, Institute for Mining and Minerals Research, 1977).

2. This strategy can be used regardless of whether an environmental damage function is specified and known. If it is unspecified, an acceptable standards policy can be used as a surrogate. See M.E. Greenbaum and C.E. Harvey, "On the Internalization of Surface Mining Externalities," *Journal of Energy Economics* vol. 2, no. 3 (July 1980), pp. 161-65.

3. The problem of uncontrolled hazardous waste disposal at the Love Canal area in upstate New York is a good example.

4. The term *bureau* is defined: "Bureaus specialize in the supply of those services that some collective organization wishes to augment beyond that supplied by the market and for which it is not prepared to contract with a profit-seeking organization." W.A. Niskanen, *Bureaucracy: Servant or Master?* (London: Institute of Economic Affairs, 1973), p. 11.

5. The term *wildcat operator* is used by the industry to describe unauthorized, illegal mining operations.

6. J. Margolis, "Shadow Prices for Incorrect or Nonexistent Market Values," in *The Analysis and Evaluation of Public Expenditures*, Subcommittee on Economy in Government (Washington, D.C.: Government Printing Office, 1969), pp. 533-46.

7. Cost-benefit and cost-effectiveness analysis are similar except that cost-benefit quantifies program outcomes in monetary terms, whereas cost-effectiveness quantifies in program terms, such as number of lives saved or number of acres reclaimed.

8. R.P. Lee and R.W. Johnson, *Public Budgeting Systems* (Baltimore: University Park Press, 1973).

9. D. McFadden, "The Revealed Preferences of a Government Bureaucracy: Theory," *Bell Journal of Economics* vol. 6, no. 2 (Autumn 1975), pp. 401-16.

10. A policy prescription should be judged by whether it achieves Pareto optimality, by whether it achieves the probable conditions suggested by positive theory, and by whether the prescription can create conditions closer to optimum than other approaches (including the policy of changing nothing). The standard Pareto condition stipulates that every individual (or homogeneous group of individuals) should be as well off as possible without any other individual (or group) being worse off.

11. A precedent for this form of organization may be found in the representative origin of the directors of the twelve U.S. federal reserve banks; three members are appointed each from industry, the public, and banking. Their function is to safeguard the financial and economic health of the nation.

VI ENVIRONMENTAL AND HUMAN ISSUES

1. OECD/IEA, *The Costs and Benefits of Sulfur Oxide Control* (Paris: OECD/IEA, 1981).

2. The category of new boilers also included existing boilers modified to lower pollution levels.

3. The bag house is a facility that receives from utility boilers the flue gas produced during the coal combustion process. The gas is routed to one of several bag house compartments, where it is directed upward into a tube-shaped bag or fabric filter.

4. W. Nesbit and R. Carr, "Particulates Caught at the Filter Line," *EPRI Journal* vol. 8, no. 7 (Sept. 1983).

5. See P. Navarro, "Our Stake in the Electric Utility's Dilemma," *Harvard Business Review*, May-June 1982.

6. The 8 million figure was subsequently increased to 10 million by a new Senate Bill 768.

7. M. Zimmerman, *The U.S. Coal Industry: The Economics of Policy Choice* (Cambridge, Mass.: MIT Press, 1981).

8. U.S. Department of Energy, Energy Information Administration, *Delays and Cancellation of Coal-Fired Generating Capacity* (Washington, D.C.: Government Printing Office, 1983).

9. J. Blodgett et al., *Acid Rain: Current Issues*, Issue Brief no. 1B 83016 (Washington, D.C.: Library of Congress, Congressional Research Service, 1984).

10. OECD/IEA, *Coal-Environmental Issues and Remedies* (Paris: OECD/IEA, 1983).

11. D. Rose, M. Miller, and C. Agnew, *Global Energy Futures and CO_2-Induced Climate Change*, MIT-EL 83-015 (Cambridge, Mass.: MIT Energy Laboratory, 1983).

12. See table 50. In 1978, one half of total output came from surface operations.

13. OECD/IEA, *Coal, Environmental Issues, and Remedies* (Paris: OECD/IEA, 1983).

14. The Office of Technology Assessment estimates that in the United States there are approximately 1,700 square kilometers of unreclaimed land. See Office of Technology Assessment, *The Direct Use of Coal*, OTA-E-86 (Washington, D.C.: Government Printing Office, 1979).

15. It is probable that the less dramatic decline in Eastern Kentucky surface mining in 1979 is due to less stringent enforcement of regulations and the vastly greater number of small, independent surface mines in operation.

16. Randall et al., *Estimating Environmental Damages from Surface Mining of Coal in Appalachia: A Case Study*, Industrial Environmental Research Laboratory (Cincinnati: U.S. Environmental Protection Agency, Office of Research and Development, 1977).

17. See W.J. Baumol, "Taxation and Control of Externalities," *American Economic Review*, June 1972.

18. Mathematica, Inc., *Design of Surface Mining Systems* (Frankfort: Kentucky Department of Natural Resources and Environmental Protection, 1974), pp. 1-30.

19. See National Safety Council, *Accident Facts* (Washington, D.C., 1968).

20. Fatality rates can be defined in a number of ways. The most concise measure is of fatalities per million labor hours worked.

21. From *Annual Reports of UMWA Welfare and Retirement Fund*, as described in G. Newmann and J.P. Nelson, "Regulation and Safety: The Effects of the Coal Mine Health and Safety Act of 1969" (Unpublished paper, June 1979), p. 26.

22. C.E. Harvey, *Financing Public Expenditures*, p. 74.

23. J. Baker, "Sources of Deep Coal Mining Productivity Change, 1962-1975," *Energy Journal*, April 1981; and W. Kruvant et al., "Sources of Productivity Decline in U.S. Coal Mining, 1972-1977," *Energy Journal*, July 1982.

VII COAL TRANSPORT AND EXPORT

1. U.S. Department of Energy, Energy Information Agency, *Coal Distribution January-December 1984*, April 1984, p. 1.

2. This procedure became known as the Ramsey Pricing Rule.

3. Association of American Railroads, *Yearbook of Railroad Facts, 1982* (Washington, D.C., 1982).

4. U.S. Department of Energy and the U.S. Departments of State and Commerce, *Report on the Potential for Cost Reduction in Inland Transportation of U.S. Coal Exports* (Washington, D.C.: Government Printing Office, 1983), p. 7.

5. U.S. Department of Energy, Energy Information Agency, *Port Deepening and User Fees: Impact on U.S. Coal Exports* (Washington, D.C.: Government Printing Office, 1983), pp. 9-10.

VIII   PROSPECTS FOR APPALACHIAN COAL

1. U.S. Congress, Joint Economic Committee, *Allocation of Resources in the Soviet Union and China—1983* (Washington, D.C.: Government Printing Office, 1984), p. 267.

2. There is evidence that suggests the OPEC price hikes of 1978-1979 reached the upper limit of world oil prices. Higher prices would not be sustainable for very long because they would encourage non-OPEC countries (those that are not members of the Organization of Petroleum Exporting Countries cartel) to increase oil production, thereby flooding the market with oil. Higher oil prices would also engender even stronger conservation measures by oil consumers and would encourage the use of substitute fuels. In fact, there is speculation that the upper limit of world oil prices now lies considerably below the even lower 1984 per barrel price of twenty-nine dollars. For a review of this, see G. Daly, J.G. Griffin, and H. Steele, "The Future of OPEC: Price Level and Cartel Stability," *Energy Journal*, January, 1983.

3. By the middle of 1983, OPEC's share in the world oil market had declined from 55% in 1974 to only 25% There is little prospect that the organization, prone to internal conflict and facing severe decline in oil demand, will reassert itself in a chronic buyer's market. In fact, in 1985 the demise of OPEC seems imminent.

4. D. Kazmer illustrates graphically the contrast between the high risk of death at conventional plants, where few people would be affected, and the low risk of death at nuclear plants, where many would be affected.

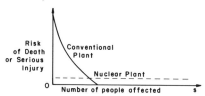

Source: David R. Kazmer, "Risk Analysis of Alternative Energy Sources," *Energy Journal* vol. 3, no. 1 (Jan. 1982), pp. 126-28.

5. M. Spangler, "Reply to 'Risk Analysis of Alternative Energy Sources,' " *Energy Journal* vol. 3, no. 1 (Jan. 1982), pp. 129-33.

6. At the end of 1982, proven reserves of U.S. natural gas were officially estimated at 200 trillion cubic feet, about a ten-year supply at current consumption rates. This is an uncomfortably low level of reserves and the lowest since estimates were first made.

7. M. Zimmerman, *The U.S. Coal Industry: The Economics of Policy Choice* (Cambridge, Mass.: MIT Press, 1981), p. 74.

8. H. Jacoby and J. Paddock, *World Oil Prices and Economic Growth in the 1980s*, MIT-EL 81-060wp (Cambridge, Mass.: MIT Energy Laboratory, 1981).

REFERENCES

Ali, M.M., C.E. Harvey, J.F. Stewart, *Regional Demand and Supply Behavior by Sectors of the U.S. Coal Industry,* Lexington: Univ. of Kentucky, Institute for Mining and Minerals Research, 1981.

Anderson, K.P., "Residential Demand for Electricity: Econometric Estimates for California and the United States," Santa Monica, Calif.: Rand Corp., 1972.

ICF Inc. and J.P. Coal Associates, *Market Guide for Steam Coal Exports from Appalachia,* prepared for Appalachian Regional Commission, Washington, D.C., March 1982.

Association of American Railroads, *Yearbook of Railroad Facts, 1982,* Washington, D.C., 1982.

Atkinson, S.E., and R. Halverson, "Interfuel Substitution in Steam Electric Power Generation," *Journal of Political Economy,* Oct. 1976.

Baker, J., "Sources of Deep Coal Mining Productivity Change, 1962-1975," *Energy Journal,* April 1981.

Battelle Memorial Institute, *Energy Perspectives,* Columbus, Ohio: Battelle, 1973.

Baughman, M.L., and F.S. Zerhoot, "Energy Consumption and Fuel Choice by Industrial Consumers in the United States," Cambridge, Mass.: MIT Energy Laboratory, March 1975.

Baughman, M.L., and P.L. Joskow, "Energy Consumption and Fuel Choices by Residential and Commercial Users in the United States," MIT Energy Laboratory, Cambridge, Mass., July 1974.

Baumol, W.J., "Taxation and Control of Externalities," *American Economic Review,* June 1972.

Bell, F. "The Pope and the Price of Fish," *American Economic Review,* December 1968.

Berndt, E.R., and D. Woods, "Technology, Price, and the Derived Demand for Energy," *Review of Economics and Statistics* vol. 57 (Aug. 1975).

Blodgett, J., et al., *Acid Rain: Current Issues,* Issue Brief 1B 83016, Washington, D.C.: Library of Congress, Congressional Research Service, Feb. 15, 1984.

Caves, R., *American Industry: Structure, Conduct, Performance,* Englewood Cliffs, N.J., Prentice-Hall, 1982.

Charles River Associates, *Coal Price Formation* (Palo Alto, Calif.: Electrical Power Research Institute, 1977).

Chase Econometrics, *Chase Energy Analysis Quarterly, Fourth Quarter, 1983,* Bala Cynwyd, Penn., Jan. 1984.

Christenson, C., and W.H. Andrews, "Coal Mine Injury Rates in Two Eras of Federal Control," *Journal of Economic Issues,* March 1973.

Committee on Science and Astronautics, *Energy Facts,* Nov. 1973.

Daly, G., J.G. Griffin, and H. Steele, "The Future of OPEC: Price Level and Cartel Stability," *Energy Journal,* Jan. 1983.

Data Resources, Inc., *Energy Review,* Winter 1983-1984, Lexington, Mass., Jan. 1984.

Edison Electric Institute, *Statistical Year Book,* 1975, New York, N.Y.

Electric Power Research Institute, *EPRI Journal,* July/Aug., 1981.

Erickson, E., and R. Spann, "Supply Response in a Regulated Industry: The Case of Natural Gas," *Bell Journal of Economics and Management Science* vol. 2 (Spring 1971), New York, N.Y.

Federal Energy Regulatory Commission (formerly Federal Power Commission), *Annual Summary of Cost and Quality of Electric Utility Plant Fuels,* 1973-1977.

Goodman, A., Unpublished paper, "The Demand for Metallurgical Coal," Johns Hopkins University, Nov. 1984.

Gordon, R.S., "A Reinterpretation of the Pure Theory of Exhaustion," *Journal of Political Economy* vol. 75 (1967).

Greenbaum, M.E., "Externalities and Acceptable Standards—The Case of Surface Mining," *Journal of Environmental Management,* July 1978.

——, *Kentucky Coal Reserves: Effects of Coal Industry Structure and Output,* Lexington: Univ. of Kentucky, Institute for Mining and Minerals Research, 1975.

Greenbaum, M.E., and C.E. Harvey. "On the Internalization of Surface Mining Externalities," *Journal of Energy Economics* vol. 2 (July 1980).

——, *Policy Alternatives for Reclaiming Surface-Mined Lands,* IRRT 42-PA1-78, Lexington: Univ. of Kentucky, Institute for Mining and Minerals Research, Dec. 1978.

Griffin, J.M., "Interfuel Substitution Possibilities: A Translog Application to Pooled Data," *International Economic Review,* Oct. 1977.

Harvey, C.E., *A Catalogue of Permit Requirements for Coal Mining and Coal Conversion Facilities in Kentucky,* Lexington: Kentucky Department of Energy, Office of Planning and Evaluation, 1979.

——, *The Economics of Kentucky Coal,* Lexington: Univ. Press of Kentucky, 1977.

——, *Financing Public Expenditures for Energy Impacted Roads,* Univ. of Kentucky, Institute for Mining and Minerals Research, November 1977.

Houthakker, H., and L. Taylor, *Consumer Demand in the U.S., 1929-1970*, Cambridge, Mass.: Harvard Univ. Press, 1966.

Hurst, E., W. Lin, and J. Cope, "An Engineering Economic Model of Residential Energy Use," Technical Report TM 5470, Oak Ridge, Tenn.: Oak Ridge National Laboratory, July 1976.

ICF, Inc., *Coal Market Assessment, Fall-Winter*, Washington, D.C., Feb. 1984.

Jacoby, H., and J. Paddock, *World Oil Prices and Economic Growth in the 1980s*, MIT-EL 81-060, Cambridge, Mass.: MIT Energy Laboratory, Dec. 1981.

Joskow, P.L., and M.L. Baughman, "The Future of the U.S. Nuclear Energy Industry," *Bell Journal of Economics*, Spring 1976.

Kalt, J.P., *Energy Supply, Environmental Protection, and Income Redistribution under Federal Regulation of Coal Strip Mining*, Studies in Energy and the American Economy, Harvard Univ. Discussion Paper no. 14, Jan. 1982; also 81-066WP (Boston, MIT, Energy Laboratory, 1982).

Kazmer, D.R., "Risk Analysis of Alternate Energy Sources," *Energy Journal* vol. 3 (Jan. 1982).

Kentucky Energy Cabinet, Division of Coal Development, "Kentucky Coal Transportation Study," Interim Report no. 2, prepared for Kentucky DOE, 1983.

Koch, J.V., *Industrial Organization and Prices*, Englewood Cliffs, N.J.: Prentice-Hall, 1974.

Kruvant, W., et al., "Sources of Productivity Decline in U.S. Coal Mining, 1972-1977," *Energy Journal*, July 1982.

LeBel, P.G., *Energy Economics and Technology*, Baltimore: Johns Hopkins Univ. Press, 1982.

Lee, R.P., and R.W. Johnson, *Public Budgeting Systems*, Baltimore: Univ. Park Press, 1973.

Man, C.E., and J.N. Heller, *Coal and Profitability: An Investor's Guide*, New York: McGraw-Hill, 1979.

Margolis, J., "Shadow Prices for Incorrect or Nonexistent Market Values," in *The Analysis and Evaluation of Public Expenditures*, Subcommittee on Economy in Government, Washington, D.C.: Government Printing Office, 1969.

Mathematica, Inc., *Design of Surface Mining Systems*, Frankfort: Kentucky Department of Natural Resources and Environmental Protection, Jan. 1974.

McFadden, D., "The Revealed Preferences of a Government Bureaucracy: Theory," *Bell Journal of Economics* vol. 6 (Autumn 1975).

Mining Informational Services, *Keystone Coal Industry Manual*, New York: McGraw-Hill, 1984.

National Coal Association, *Long Range Forecast for U.S. Coal, Coal Markets in the Future*, Washington, D.C., March 1984.

————, *Steam Electric Plant Factors*, 1966, 1972, 1977, 1981, Washington, D.C..

National Safety Council, *Accident Facts*, Washington, D.C., 1968.

Navarro, P., "Our Stake in the Electric Utility's Dilemma," *Harvard Business Review*, May-June 1982.

Nerco Coal Company, *General Marketing Trends and Forecasting*, Cincinnati, July 1984.

Nesbit, W., and R. Carr, "Particulates Caught at the Filter Line," *EPRI Journal* vol. 8 (Sept., 1983).

Newmann, G., and J.P. Nelson, "Regulation and Safety: The Effects of the Coal Mine Health and Safety Act of 1969," unpublished paper, June 1979.

Niskanen, W.A., *Bureaucracy and Representative Government*, Chicago: Aldine-Atherton, 1971.

———, *Bureaucracy: Servant or Master?* London: Institute of Economic Affairs, 1973.

Organization for Economic Cooperation and Development, International Energy Agency, Paris, France: OECD/IEA.

———, *Coal: Environmental Issues and Remedies*, 1983.

———, *1982 Coal Information Report*, 1982.

———, *Coal Prospects and Policies, 1983 Review*, 1984.

———, *The Costs and Benefits of Sulfur Oxide Control*, 1981.

———, *Steam Coal*, 1978.

———, *World Energy Outlook*, 1982.

Perry, C., *Safety Laws and Spending Save Lives: Analysis of Coal Mine Fatality Rates 1930-1979*, RS-69, Lexington: Univ. of Kentucky, Department of Sociology, Dec. 1981.

Perry, C., and C. Ritter, *Dying to Dig Coal: Fatalities in Deep and Surface Coal Mining in Appalachian States, 1930-1978*, RS-68, Lexington: Univ. of Kentucky, Department of Sociology, Dec. 1981.

Pindyck, R.S., "The Optimal Production of an Exhaustible Resource When Price is Exogenous and Stochastic," *Scandinavian Journal of Economics* vol. 83 (1981).

———, *The Structure of World Energy Demand*, Cambridge, Mass.: MIT Press, 1979.

———, "Uncertainty and Exhaustible Resource Markets," *Journal of Political Economy* vol. 88 (1980).

Randall, et al., *Estimating Environmental Damage from Surface Mining of Coal in Appalachia: A Case Study*, Cincinnati: U.S. Environmental Protection Agency, Office of Research and Development, Industrial Environmental Research Laboratory, 1977.

Rose, D., M. Miller, and C. Agnew, *Global Energy Futures and CO_2 Induced Climate Change*, MIT-EL 83-015, Cambridge, Mass.: MIT Energy Laboratory, 1983.

Schweitzer, A., *The Limits of Kentucky Coal Output: A Short Term Analysis*, Lexington: Univ. of Kentucky, Institute for Mining and Minerals Research, 1973.

Shrieves, R.E., "Geographic Market Areas and Market Structure in the Bituminous Coal Industry," *The Antitrust Bulletin* vol. 23 (Fall 1978).

Spangler, M. "Reply to 'Risk Analysis of Alternate Energy Sources,' " *Energy Journal* vol. 3 (Jan. 1983).

Stigler, G., *The Theory of Price,* New York: Macmillan, 1966.

U.S. Congress, House, 93 Cong. 1 sess., H.R. 8619, P.L. 93-135, Washington, D.C., 1973.

U.S. Congress, Joint Economic Committee, *Allocation of Resources in the Soviet Union and China-1983,* Washington, D.C.: Government Printing Office, 1984.

U.S. Congress, Senate Committee on Interior and Insular Affairs, 93 Cong., *Coal Surface Mining and Reclamation,* Washington, D.C., 1973.

U.S. Department of Commerce, Bureau of Census, *Monthly Report,* EM 522.

U.S. Department of Energy and Departments of State and Commerce, *Report on the Potential for Cost Reduction in Inland Transportation of U.S. Coal Exports,* Washington, D.C.: Government Printing Office, Aug. 1983.

U.S. Department of Energy, Energy Information Administration, Washington, D.C.: Government Printing Office.

———, *Annual Energy Review,* April 1983, April 1984.

———, *Coal Distribution,* 1977-1984.

———, *Coal Production,* 1980-1984.

———, *Coke and Coal Chemicals,* 1977-1980.

———, *Cost and Quality of Fuels for Electric Utility Plants,* 1978-1982.

———, *Delays and Cancellations of Coal-Fired Generating Capacity,* July 1983.

———, *Demonstrated Reserve Base of Coal in the United States on Jan. 1, 1980,* May 1982.

———, *Impacts of the Proposed Clean Air Act Amendments of 1982 on the Coal and Electric Utility Industries,* June 1983.

———, *Interim Report of the Interagency Task Force,* 1981.

———, *Monthly Energy Review,* April 1982, Sept. 1982, Feb. 1984, March 1985.

———, *Port Deepening and User Fees: Impact on U.S. Coal Exports,* May 1983.

———, *Quarterly Coal Report,,* April 1983, April 1984, July 1984.

———, *Railroad Deregulation: Impact on Coal,* Aug. 1983.

———, *Weekly Coal Production,* Nov. 6, 1982; Jan. 21, 1984; Jan. 29, 1983; May 6, 1983; March 2, 1984.

U.S. Department of Interior, Bureau of Mines, Washington, D.C.: Government Printing Office.

———, *Demonstrated Coal Reserve Base,* May 1973.

———, *Minerals Industry Surveys,* 1970-1973.

———, *Minerals Yearbook,* 1970-1976.

U.S. Department of Labor, Mine Safety and Health Administration, *Injury Experience in Coal Mining,* Washington, D.C., 1984.

U.S. Federal Trade Commission, *In the Matter of Kennecott Copper Corp., Findings of Fact, Conclusion and Final Order,* Docket 8765, Washington, D.C., May 5, 1971.

United Mine Workers of America, Welfare and Retirement Fund, *Annual Reports.*

Walton, D.R., and P.W. Kaufman, *Preliminary Analysis of the Probable Causes of Decreased Coal Mining Productivity* (1969-1976), Management Engineers, Inc., for U.S. Department of Energy, Division of Solid Fuels Mining and Preparation, Nov. 1977.

White, L., "Searching for the Critical Industrial Concentration Ratio" in *Studies in Nonlinear Estimation,* ed. S. Goldfeld and R. Quandt, Cambridge, Mass.: Ballinger, 1976.

Wilson, C.L., *Coal: Bridge to Future,* Cambridge, Mass.: Ballinger, 1980.

Zimmerman, M., *The U.S. Coal Industry: The Economics of Policy Choice,* Cambridge, Mass.: MIT Press, 1981.

INDEX